Composites: Mechanical, Physical and Other Properties

Composites: Mechanical, Physical and Other Properties

Edited by **Gerald Brooks**

New York

Published by NY Research Press,
23 West, 55th Street, Suite 816,
New York, NY 10019, USA
www.nyresearchpress.com

Composites: Mechanical, Physical and Other Properties
Edited by Gerald Brooks

© 2015 NY Research Press

International Standard Book Number: 978-1-63238-086-9 (Hardback)

Printed in the United States of America.

Contents

Preface   VII

Section 1   Mechanical and Physical Properties of Composites   1

Chapter 1   **Characterizations of Environmental Composites**   3
Ali Hammood and Zainab Radeef

Chapter 2   **Friction and Wear of Polymer and Composites**   21
Dewan Muhammad Nuruzzaman
and Mohammad Asaduzzaman Chowdhury

Chapter 3   **Frequency-Dependent Effective Material Parameters
of Composites as a Function of Inclusion Shape**   53
Konstantin N. Rozanov, Marina Y. Koledintseva
and Eugene P. Yelsukov

Chapter 4   **The Chosen Aspects of Materials
and Construction Influence on the Tire Safety**   81
Pavel Koštial, Jan Krmela, Karel Frydrýšek and Ivan Ružiak

Chapter 5   **The Lightweight Composite Structure and
Mechanical Properties of the Beetle Forewing**   115
Jinxiang Chen, Qing-Qing Ni and Juan Xie

Chapter 6   **Comparative Review Study on Elastic Properties
Modeling for Unidirectional Composite Materials**   147
Rafic Younes, Ali Hallal, Farouk Fardoun and Fadi Hajj Chehade

Section 2   Metal and Ceramic Matrix Composites   165

Chapter 7   **Manufacturing and Properties of Quartz (SiO_2)
Particulate Reinforced Al-11.8%Si Matrix Composites**   167
M. Sayuti, S. Sulaiman, T.R. Vijayaram,
B.T.H.T Baharudin and M.K.A. Arifin

Chapter 8   **YSZ Reinforced Ni-P Composite**
             by Electroless Nickel Co-Deposition   193
             Nor Bahiyah Baba

Chapter 9   **Carbon Nanotube Reinforced**
             Alumina Composite Materials   219
             Go Yamamoto and Toshiyuki Hashida

Chapter 10   **Characterisation of Aluminium Matrix Syntactic Foams**
             Under Static and Dynamic Loading   239
             M. Altenaiji, G.K. Schleyer and Y.Y. Zhao

Permissions

List of Contributors

Preface

In my initial years as a student, I used to run to the library at every possible instance to grab a book and learn something new. Books were my primary source of knowledge and I would not have come such a long way without all that I learnt from them. Thus, when I was approached to edit this book; I became understandably nostalgic. It was an absolute honor to be considered worthy of guiding the current generation as well as those to come. I put all my knowledge and hard work into making this book most beneficial for its readers.

Composite materials can be utilized in various industries such as aerospace and construction industries as they are a class of materials which possess some unique and extraordinary properties. Recently, composites have received a lot of attention due to their role in the field of material research and also because of the advent of new forms of composites such as nanocomposites and bio-medical composites. This book primarily deals with the production and property classification of various composites. It covers two broad sections: Mechanical and Physical Properties of Composites, and Metal and Ceramic Matrix Composites.

I wish to thank my publisher for supporting me at every step. I would also like to thank all the authors who have contributed their researches in this book. I hope this book will be a valuable contribution to the progress of the field.

Editor

Mechanical and Physical Properties of Composites

Characterizations of Environmental Composites

Ali Hammood and Zainab Radeef

Additional information is available at the end of the chapter

1. Introduction

Recently, environmental preservation issues have been critical between the chemical pollution matters and the development technology requirements. However, the renewable and friendly materials come to use.

Numerous researches have richly studies the natural fiber reinforcement polymer composites. This fact, based on both fibers and matrixes are derived from renewable resources. Therefore, the formed composites have more compatibility with the environmental preservation issues. [1] Isabel investigated of the most natural fibers are used such as palm, cotton, silk, coconut, wool and wood fibers. A significant development in the lignocelluloses fiber in thermoplastics realized the distinct researches presented by [2-5] Composite-reinforcing fibers can be categorized by chemical composition, structural morphology, and commercial function. Natural fibers, such as kenaf, ramie, jute, flax, sisal, sun hemp and coir are derived from plants that used almost exclusively in PMCs. Aramid fibers [6] are crystalline polymer fibers are mostly used to reinforce PMCs. The compounds percentage of composite have the essential role for verify the designed values according to applications, therefore the mechanical properties of PMCs predicated by Mohamed (2007).

The primary function of a reinforcing fiber is to increase the strength and stiffness of a matrix material. The fibers reinforced composite have the essential role in this investigation for its significant property advantages as high stiffness, lightweight, easily recycled material, availability, low manufacturing cost, the environment effect and lifetime rupture behavior. Various types of natural fibers are available to combine with other mineral fiber for construct composite material. Essentially, the fiber can be classified as vegetable, animal, and man-made fibers. The main disadvantages of natural fibers are their high level of moisture absorption, poor and interfacial adhesion, relatively low heat resistance. [7-8] investigated high speed impact events using (PKV, PRM) composites. This research was indicated significant improvements in the penetration resistance. This fact comes from the improvement of target geometry structure. Numerous researches have been carried out on the ballistic impact on high strength fabric structures [9-11]. In the airport and marine applications, the dynamic loads effect and the chemo interactions were attracted the researchers and many methods are employed for computing the surface topography parameters, thereby numerous estimations were covered the erosion-corrosion behavior of PMCs [12-13]. The most impinging parameters focused on environment effects and impinging angle [14-17]. There were many instruments and electronic microscopes developed with the time for measuring the roughness parameters and drawing the surface topography [19]. The Erosion and corrosion of composite must be determent for the accelerator objects, whereby this values will be indicator for measuring the life time rupture of composite [18].

2. Important

The automotive and aerospace industries have both shifted for using natural fiber reinforced composites as a factor to reduce the weight and getting significant properties of composite components. As matter of fact, the impinging liquids of the naval and aerospace applications have a direct effect on surface topography. Therefore, advanced studies focused on corrosion and erosion behavior. The impinging angle, velocity versus time, composite morphology represented the essential parameters of this field of study. In this investigation, surface roughness versus time was the indicator for erosion and corrosion effects.

3. Experiment Procedure and Samples Preparation

In order to develop new composite material with high impact resistance and high erosion resistance, characterization study for two sets of composites materials have been computed. The specific composites materials in this research are: polyester resin-matrix and Kevlar reinforced fiber (PKV) with V_f (42%), polyester resin–matrix and ramie reinforced fiber (PRM) with V_f (42%). Experimental program was carried out to study the erosion and corrosion behavior by computing the surface roughness parameters of (PKV, PRM) before and after impinging operation. Hence, Polymer Matrix Composites (PMCs) were examined by impingement using water jets when the aqueous solution was 3.5 wt% NaCl. Erosion and corrosion tests were impinging at 90 angles at velocity 30 m/s and the impinging period was 12 hours.

The composite subject study consists from five Kevlar layers and five ramie layers were impregnated with unsaturated polyester. The layers aligned alternatively according to the expectation performance. The synthetic ramie fiber was weaved as shown in Figure 1(a) & (b).

4. Tensile testing of composites

Tensile specimen (145 x 15 x 7) mm3 are caught according to ASTM D 3039 /D 3039M-95M standard (ASTME D 3039, 2003). Tensile test has been conducted and the data acquired digitally. Tensile stress, tensile strain and Young modulus of PKV with Vf (42%) and PRM with Vf (42%) were calculated and the test was performed by instron machine 10 KN, series

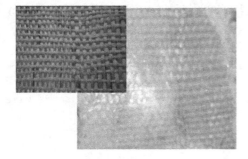

(a) Ramie Woven and PRM   (b) Kevlar Woven

Figure 1. Ramie and Kevlar Woven

2716 and 2736 under stable speed rate 2 mm/min. The test has verified all the specific test conditions to determine the tensile properties of specimen according to the ISO 527. Tensile test has been performed to estimate the yield stress, young modulus and tensile extension at yield point. Additionally, Poisson's ratio has been calculated by evaluate transverse strain and longitudinal strain of composite. Hence, the transverse strain calculated by using strain gage that supplied from Tokyo Sokki Kenkyuio co, Ltd. The used gage type was BFLA-2-8 with gage resistance 120±0.3 Ω. Then, the strain is connected to DAQ bridge system for reading the transverse strain.

The composite behavior under test was an important subject for this investigation and the specimen geometry was constructed from five layers Ramie fiber and five layers of Kevlar as clarified in the Figure 2.

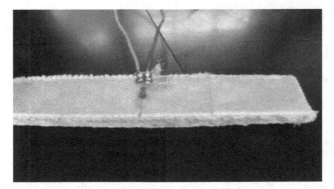

Figure 2. Tensile Test Specimen with Strain Gage

5. Ballistic limit and number of layers

The ballistic limit is commonly defined as a 50% probability of penetrating a target at a given impact velocity. The energy absorption is related to the impact velocity, interpreted by the effect of the striking velocity on the amount of kinetic energy that is absorbed by the composite material. Hence, the energy absorbed by the fabric is equal to the residual energy amount subtracted from total impact energy.

In this test, the target impinged by using gas gun machine supported by a high speed camera for record the impact event at 30,000 frames per second with image size 512 x 64 pixels per image [7]. Figure 3 (a) shown the gas gun machine. Thus, The velocity before the target and the residual velocity after the target were estimated. All the targets were impinged by Semi-conical bullets as shown in the Figure 3(b).

(a) Ballistic Panel   (b) Semi-conical bullets

Figure 3. Gus gun device of University Putra Malaysia [7,8]

6. Erosion test and corrosion

Accurate estimation is carried out for calculating the erosion and corrosion percentage of (PKV, PRM) samples. The tests were conducted using a jet at velocity 30 m/s and impinging angles 90°. Essentially, the exposure period was the major factor of erosion estimation. Hence, the impinging periods were from 3 to 6 and from 6 to 12 hours under room temperature. Samples roughness data were recorded before and after the tests by Image processing software for scanning probe microscope

7. Result and discussions

7.1. Tensile test

The tensile test of this composite was conducted for specifying the mechanical properties of the composite. Generally, the test results recorded high tensile strength. The brittle manner was the first stage of composite under test due to the low elongation ability of matrix. The second stage was the ductile behavior that embodied of high elongation ability of Kevlar layer. From stress – strain curve topography, the specimen have extension at maximum load up to 2.49 mm was recorded. There are continuously reductive of the curve as a result to the elongation of Kevlar layers. The extension at maximum tensile strain observed at 43.25 mm that was evident in the ductile behavior at front face of the specimen. Practically, through the extension test, the back face of the specimen that contented from ramie layers reinforced polyester was separated gradually as result to the brittle behavior of ramie – polyester matrix. Automatic Young's modulus in this composite is 4930.5 MPa. This fact plotted in stress – strain curve in the Figure 4. The composite properties depend at compounds types and volume fraction. Therefore, the volume fraction of fiber or matrix could be calculated as it is clarified in the following equation.

$$V_m = \frac{v_m}{vc}$$ (1)

Where
V_m = volume fraction of matrix
v_m = volume of matrix
v_c = total volume of the composite

Volume fraction of fiber could be evaluated from.

$$V_m = 1 - V_f$$ (2)

Where
V_f = volume fraction of fiber

The volume of ramie fiber could be calculated from the following equation.

$$V_f = V_K + V_R$$ (3)

Where
V_k = volume fraction of Kevlar
V_R = volume fraction of ramie

Fiber density can be determined experimentally by weighting the Kevlar and ramie fibers and calculated the volume fraction for the fibers and resin. Therefore, from the following expression could be calculated the density.

$$\delta_f = \frac{\delta_c - V_m \delta_m}{V_f}$$ (4)

Where
δ_f = Fiber density
δ_c = Composite density
δ_m = Matrix density

The Poisson coefficient represents the contraction in the transverse direction and could be calculated by using the follow expression.

$$V_{lt} = v_f V_f + v_m V_m \tag{5}$$

$$V_{lt} = 0.34 \tag{6}$$

V_{lt} = Poisson ratio

Where the Kevlar Poisson ratio is equivalent to 0.34 [6], ramie Poisson ratio is equal to 0.3 [20] and 0.4 for unsaturated polyester resin [21].

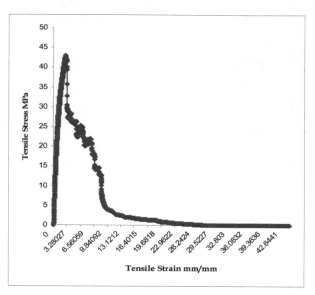

Figure 4. Stress – Strain Curve of PKV +PRM Composite

Modulus of elasticity (E_{11}) can be calculated in (7).

$$E_{11} = \frac{\sigma_L}{\varepsilon} \tag{7}$$

σ_L =longitudinal tensile stress
ε =strain

(E_{22}) is calculated by using gage strain that recorded the shrinking displacement and more than 550 data point recorded load and displacement. The transverse Young's Modulus is the initial slope of σ tra. ε_2 curve.

The Young's modulus can be calculated by using the retrieved data from Figure 5 as clarified in the following equation.

$$E_{22} = \frac{\sigma_{tra}}{\varepsilon_2} \tag{8}$$

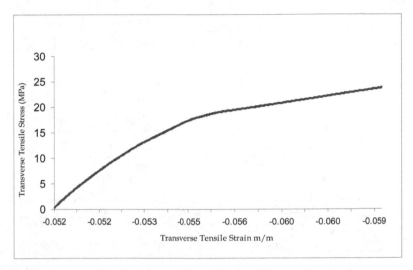

Figure 5. Stress – Transverse Tensile Stress and Strain Curve of Composite

Where

E_{22} = Transverse Young's Modulus

σ_{tra} =Transverse tensile stress

ε_2 = Transverse strain

The maximum shear stress can be obtained from the following equation.

$$\tau_{max} = \frac{\sigma_y}{2}$$ (9)

Where

σ_y = Stress in the yield point

τ_{max} = Maximum shear stress

Shear modulus (G_{12}) can be calculated from the load data that plotted in the shear stress curve.

$$G_{12} = \frac{\tau}{\gamma}$$ (10)

Where

τ = shear stress

γ = shear strain

In addition, the Poisson ratio can be estimated from the following equation and the result seems equivalent to the Poisson ratio that was calculated according to the compounds volumes fraction. Table 1 presents the longitudinal and transverse averages of data.

$$\gamma_{12} = \frac{\varepsilon_{lat}}{\varepsilon_{long}} \qquad (11)$$

Where

ε_{lat} =Transverse strain

ε_{long} =Longitudinal strain

The average of longitudinal data for Kevlar – Ramie –Polyester composite				
E_{11} (GPa)	$(\sigma_1)_{ult}$ (MPa)	$(\varepsilon_1)_{ult}$	ζ_{max} (MPa)	Poisson's Ratio
3.9489±0.5	66.75±5.4	0.125±0.04	33.37±3	0.37

The average of transverse data for Kevlar – Ramie –Polyester composite				
E_{22} (MPa)	$(\sigma_2)_{ult}$ (MPa)	$(\varepsilon_2)_{ult}$	γ_{12}	G_{12} (MPa)
244.74±2.5	22.77±1.3	0.0735±0.03	0.257±0.05	132.65±15

Table 1. Tensile Test Data

7.2. High speed impact results

Understanding the impact response of composites has be come an area of great academic and practical interest. The major advantages of composite materials are their high strength and stiffness, light weight, corrosion resistance, crack, fatigue resistance and flexibility. Ramie – Kevlar reinforced polyester resin present high resistance. The high level from resistance could be realized by increasing of Kevlar layers.

$$[E_{abs} = \frac{1}{2}m \ (V_{imp}^2 - V_{res}^2)] \qquad (12)$$

Where

E_{abs} = Energy absorption

m = projectile mass

V_{imp} = Impact velocity

V_{res} = Residual velocity

Ballistic limit can be identified as the limit between the penetration and the fully arrested. Thus, the composite with five layers (PKV) and five layers (PRM) couldn't meet the specific requirements of ballistic resistance. In fact, the absorption of energy will be increased with increase the number of layers [7, 8].

In the event when no perforation occurs, the energy absorbed by the target will be equal to the initial impact energy. The high speed impact data for PKV & PRM are shown in the table 2. The following equation has been verified to the ballistic limit or fully arrested action.

$$[E_{abs} = \frac{1}{2} m \ V_{imp}^2]$$ (13)

Where

E_{abs} = absorption of energy

V_{imp} = impact velocity

In this prospective must be mention the most high speed impact parameters that represented rich fields of studies are, target geometry, projectile type, target thickness, composite compounds.

Humanity:53%	Bullet type: Semi-conical
Specimen type: TSP	Camera temperature:40 °C
Target area:15 × 10 mm	Temperature: 32 °C
Material: PKV& PRM	Camera resolution: 512 × 48

Layer No.	Gas gun Pressure Psi	Initiation Velocity m/s	Residual velocity m/s	Absorption of Energy J
	300	273.9	125.9	147.926
5K-5R	250	275.255	150.17	133.035
	250	255.07	120.9	126.109

Table 2. High Speed Impact of PKV & PRM

7.3. Erosion result

The surface roughness of engineering applications has interacted with the environment. Therefore, the studies pay attention for estimating the surfaces roughness of materials with respect to essential parameters to limit the wear mechanism of materials. Roughness value can either be calculated on a profile or on a surface Rz, Rq, and Sa is the arithmetic average of the 3D roughness. Hence, the impinging test conduct through specific periods is illustrated in the Table (3).

Kevlar roughness by (nm)				Ramie roughness by (nm)			
0 hours	3 hours	6 hours	12 hours	0 hours	3 hours	6 hours	12 hours
8.92	11	9.36	9.94	12.9	11	6.71	10.8
16.7	8.25	8.06	10.4	7.37	9.82	6.85	10.4
15.6	8.44	7.94	11	10	8.32	7.78	10.3
7.53	9.26	6.35	8.78	9.12	11	7.82	8.43
6.06	8.69	6.24	11.6	10.6	10.9	8.25	8.8
6.44	12.9	9.45	10.4	11.3	8.31	9.23	9.18
6.28	12.9	8.81	14	10.6	11.3	9.97	9.57

Table 3. Roughness values

The roughness parameters rates, such as (Amplitude parameters, Hybrid Parameters and Functional Parameters) were estimated by all morphology images with image size (100.00nm X 80.00nm) Figure 6 illustrate the morphology images before PKV tests.

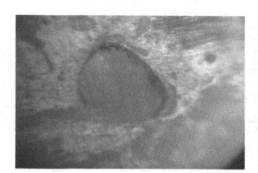 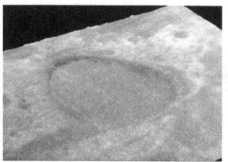

Figure 6. 3D Morphology Images after Tests of PKV.

The erosion wear loss was determined using probe microscope. Typically, scanning probe microscope image process software provided reliable means to evaluate the erosion volume lose. This technique characterizes and quantifies the surface roughness parameters, the surface profile and topographical features in three-dimension using high precision range observed at 100 nm. All measurements were made with an effective magnification of X 12.5. Excellent data were recorded for average size and average height for surface grains. Hence, the average volume of grains can be computed by the grain size analyzer. Randomly, erosion areas were elected for scanning morphologic image. The erosion volume loss was derived from analyzing the erosion surface at three dimensions. Therefore, the erosion volume loss (Vloss) can be expressed as the average volume of surface at zero erosion time (V0) subtracted from the average volume of surface after specific time (Vt). Eroded area was randomly measured at seven locations. Then, the average of erosion volume was calculated.

$$V_{loss} = V_0 - V_T \tag{14}$$

According to the V_{loss} formula, the averages of erosion volume loss were illustrated in Figure 7.

Ramie reinforced polyester matrix present higher value of volume erosion loss than Kevlar reinforced polyester. This fact derives from the poor of adhesion between Kevlar filaments and polyester resin. Table 4 illustrates the grains size versus time.

volume loss	Kevlar			Ramie		
	avg. size (nm²)	avg. height (nm)	avg. volume	avg. size (nm²)	avg. height (nm)	avg. volume
Zero Time	1.13	7.14	8.21	1.24	8.97	11.18
after 3 hours	1.11	10.334	10.44	0.82	9.36	7.846
after 6 hou	1.72	8.91	11.55	0.9	5.41	4.25.003

Table 4. Grains size versus time

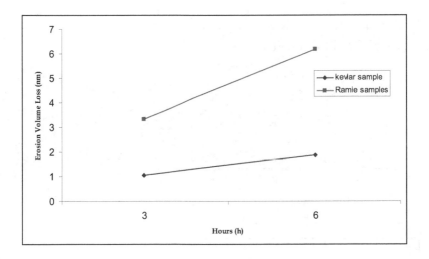

Figure 7. Erosion volume loss concomitant with time

Transforming image to 3D topography with 200% zoom area for PKV sample at zero time (before test) illustrated accurate 3D surface profile with 243 peaks number and the maximum height 46 nm as shown in the Figure 8 and Figure 9 shown the PKV after 12 hours illustrated in 3D surface with 238 peaks number and maximum height 73.6 nm.

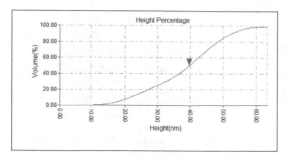

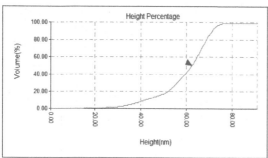

Figure 8. PKV at Zero Time (before test) in 3D Image

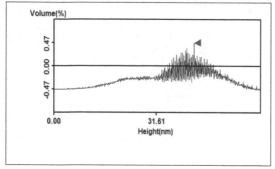

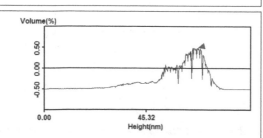

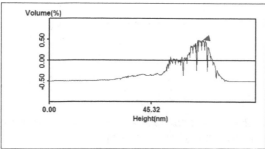

Figure 9. PKV after 12 Hours in 3D Image

Accurate roughness parameters were established for PKV and PRM versus the time before and after test as illustrates in the table 5. PKV & PRM morphologies images versus time were illustrated in Figure 10 & 11.

Image size:100.00nm X 80.00nm (PKV) at zero time					
Amplitude parameters:	Area 1	Area 2	Area 3	Area 4	Area 4
Sa(Roughness Average) [nm]	6.06	7.53	8.92	15.6	16.7
Sq(Root Mean Square) [nm]	9.1	10.7	12.3	18.8	20.2
Ssk(Surface Skewness)	-2.32	-1.67	-1.58	-0.451	-0.386
Sku(Surface Kurtosis)	12.7	8.92	6.75	2.41	2.44
Sy(Peak-Peak) [nm	87.8	90.2	89.5	90.3	91.4
Sz(Ten Point Height) [nm]	85.2	89	85.8	89.9	88.5
Hybrid Parameters:					
Ssc(Mean Summit Curvature) [1/nm]	-155	-161	-146	-146	-149
Sdq(Root Mean Square Slope) [1/nm]	10.1	10.7	8.12	8.69	8.02
Sdr(Surface Area Ratio)	3.29E+03	3.62E+03	2.17E+03	2.44E+03	2.08E+03
Functional Parameters:					
Sbi(Surface Bearing Index)	0.624	1.05	0.779	2.45	14.1
Sci(Core Fluid Retention Index)	0.983	1.18	0.949	1.34	1.32
Svi(Valley Fluid Retention Index)	0.192	0.17	0.211	0.103	0.112
Spk(Reduced Summit Height) [nm]	4.79	7.41	4.61	7.98	0.355
Sk(Core Roughness Depth) [nm]	16.4	20.1	22.4	44.4	55.1
Svk(Reduced Valley Depth) [nm]	21.5	20.8	25.2	19.9	19.6
Sdc 0-5(0-5% height intervals of Bearing Curve) [nm]	14.6	10.2	15.7	7.67	1.43
Sdc 5-10(5-10% height intervals of Bearing Curve)[nm]	1.72	3.08	1.31	4.32	0.803
Sdc 10-50(10-50% height intervals of Bearing Curve) [nm]	7.63	9.69	10.3	19.4	26.9
Sdc 50-95(50-95% height intervals of Bearing Curve) [nm]	16.6	19.9	26	36.1	35.7
Spatial Parameters:					
Sds(Density of Summits) [1/um2]	4.28E+06	4.23E+06	6.02E+06	5.67E+06	5.56E+06
Fractal Dimension	3	3	2.81	2.37	2

Table 5. The roughness parameters rates

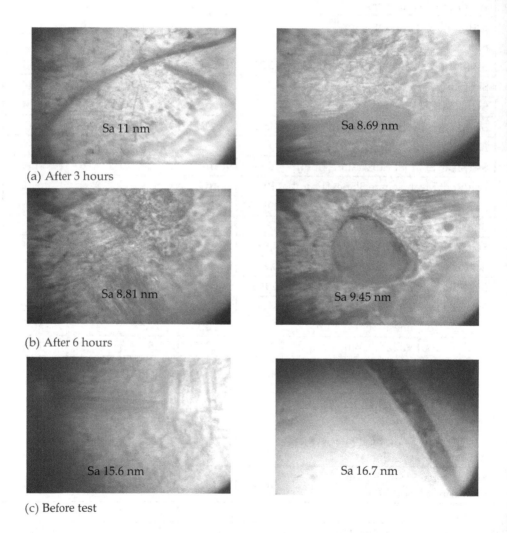

(a) After 3 hours

(b) After 6 hours

(c) Before test

Figure 10. Erosion and corrosion test for Kevlar – polyester composite morphology and Sa(Roughness Average)

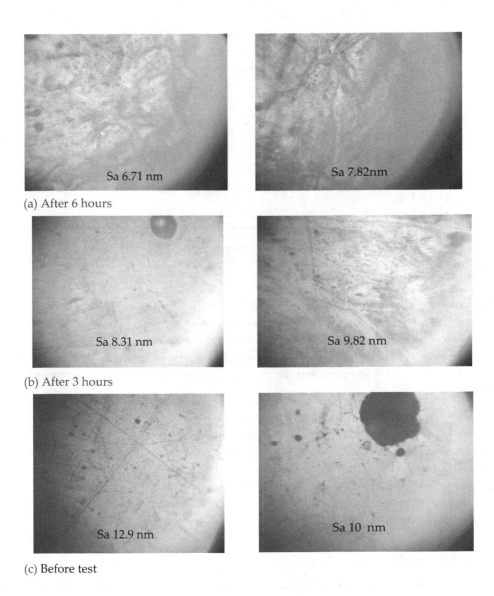

(a) After 6 hours

(b) After 3 hours

(c) Before test

Figure 11. Ramie – polyester composite morphology and Sa(Roughness Average) after erosion-corrosion test

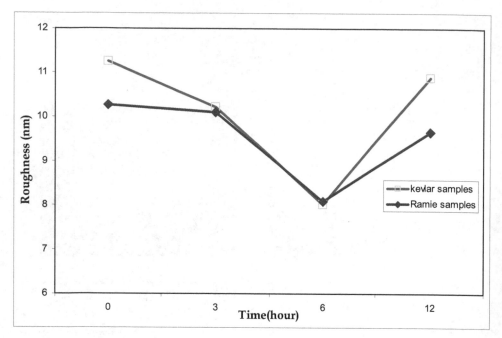

Figure 12. The Roughness Rates Versus Time

The roughness rate of the Kevlar–polyester and ramie-polyester concomitant time tented to be equivalent values. Really, the polyester reinforced faced the impinged water has the major role in the erosion and corrosion resistance. The Kevlar and ramie fibers have specific interfacial adhesion with the matrix. The ramie fibers presented high interfacial adhesion with resin as a result to plant fiber nature. On the other hand, the Kevlar presented a poor linkage with polyester resin.

All the fiber layers were protected by the matrix mass. Figure12 illustrated a declining in the polyester surface roughness versus time. The roughness will be the function for estimating the erosion rate. Therefore, within the period, from 3 until 6 hours the impinging effect assisted of soften the surface, but within period from 6 until 12 hours the erosion rate will be increase and the surface tent to be rough. Electrochemical test have been conducted and there are no significant corrosion data recorded, this finding derived from the compound nature of composite.

8. Conclusions

Generally, plain weave represented the most common fabric, due to the significant properties that embodied their tensile strength to weight ratio. The study of the stress – strain response, high speed impact evaluation, and erosion corrosion behavior leads to the following conclusion:

1. The Kevlar –polyester composites behave ductile manner but the ramie-polyester composites behave brittle manner.
2. The roughness rate of the Kevlar-polyester and ramie-polyester decreases through 3 -6 hours but there is an increase in the roughness rate through 6-12 hours due to the increasing in erosion rate of polyester matrix.
3. All roughness parameter computed accurately.
4. PKV and PRM with five layers PKV and five layers PRM present fully penetration.
5. No significant corrosion has been recorded.

Author details

Ali Hammood and Zainab Radeef
Department of Materials Engineering-University of Kufa, Iraq

Acknowledgement

The authors would like to express their gratitude and sincere appreciation to the department of Mechanical and Manufacturing Engineering of the University Putra Malaysia and Material Engineering Department- College of Engineering-University of Kufa for scientific assistance and support.

9. References

[1] Isabel B Wingate (1976) Textile fabrics and their selection. Library of congress cataloging in publication data.
[2] Woodhams R T, Thomas G, Rodgers D K (1984) wood Fiber as Reinforcing Fillers forpolyolefins.polym.eng.Sci. Vol(24):1166.
[3] Klason C, Kubat J (1989) Cellulose in polymer composites. In composite Systems from Natural and synthetic polymers, Salmen L, de Ruvo, A Sefe-ris, J C Stark, E B Eds; Elsevier Science: Amsterdam
[4] Myers G E, Clemons C M, Balatinecz J J, Woodhams R T(1992) Effects of composite and polypropylene Melt Flow on polypropylene-Waste Newspaper Composites. proceedings on the Annual Technical Coneference; Society of plastics Industry p 602
[5] Kokta, B V, Raj R. G, Daneault C (1989) Use of Wood Flour as Filler in Polypropylene; Studies on Mechanical Properties. Polymplast. Tecnol.Eng. 28, 247.
[6] Yang H H (1993) Kevlar Aramid Faiber. west Sussex PO191UD,England.
[7] Aidy A, Shaker Z R, Kahalina A (2010) Development of anti-ballistic board, fiber polymer-plastics technology and engineering VOL.(50):622-634.
[8] Zainab Shaker Radif, Aidy Ali & Khalina Abdan(2010) DEVELOPMENT OF A GREEN combat armour from rame-kevlar- polyester composite, Pertanika Journal of Science and Technology. Vol(19) : PP 339-348,.
[9] Lee B L, Song J W, Ward J E J (1994) Compos Mater.VOL (28):PP 1202 – 1226.

[10] Goldsmith W, Dharan CK.,Chang H (1995) Quasi-static and ballistic perforation of carbon fiber laminates .Int J Solid Struct VOL(32):89-103.

[11] Almohandes AA, Abdel – Kader MS, Eleiche, AM (1996) Experimental investigation of the ballistic resistance of steel-fiberglass reinforced polyester laminated plates. Composites: Part B 27:447-58.

[12] Bardal E, Eggen T G, Rogne T, Solem T (1995) The erosion and corrosion properties of thermal spray and other coatings, in: Proceedings of the Int. Therm. Spray. Conf., Kobe, Japan.

[13] Puget Y, Tretheway K R,Wood R J K (1998) The performance of cost effective coatings in aggressive saline environments, NACE Corrosion PP 688.

[14] Burstein G T, Sasaki K (2000) Effect of impact angle on the slurry erosion–corrosion of 304 L stainless steel, Wear VOL (240): 80–94.

[15] Dawson J L, Shih C C, John C C, Eden D A (1987), Electrochemical testing of differential flow induced corrosion using jet impingement rigs, NACE Corrosion, Paper no. 453,.

[16] Clark H M, Wong K K, Impact angle (1995), particle energy and mass loss in erosion by dilute slurries, Wear 186–187 454–464.

[17] Stack M M, Zhou S, Newman R C, Identifications of transitions in erosion–corrosion regimes in aqueous environments, Wear 186 (1995) 523–532.

[18] Sherrington.(1988), modern measurement techniques in surface metrology, wear VOL(125):271-288.

[19] Matsuno Y., Yamada H., Harada M. and Kobayashi A. (1975), The microtopography of the grinding wheel surface with SEM, Ann.CIRP VOL(24):PP 237-242.

[20] Nakamura, Y et al (2003). Neurosci. Abst. 608.5

[21] Girardi M. A and Phill M. G. (1993), Microstructure and properties of polyester/urethane acrylate thermosetting blends, and their use as composite matrices, Journal of Materials Science, Volume 28. 3116-3124, DOI: 10.1007/BF00354718.

[22] Mohamed Thariq. (2007), High velocity Impact analysis of glass epoxy-laminate plates. Thesis, university Putra, Malaysia, Malaysia.

Chapter 2

Friction and Wear of Polymer and Composites

Dewan Muhammad Nuruzzaman and Mohammad Asaduzzaman Chowdhury

Additional information is available at the end of the chapter

1. Introduction

Polymer and its composites are finding ever increasing usage for numerous industrial applications in sliding/rolling components such as bearings, rollers, seals, gears, cams, wheels, piston rings, transmission belts, grinding mills and clutches where their self lubricating properties are exploited to avoid the need for oil or grease lubrication with its attendant problems of contamination [1]. However, when the contact between sliding pairs is present, there is the problem of friction and wear. Yamaguchi [2], Hooke et al. [3] and Lawrence and Stolarski [4] reported that the friction coefficient can, generally, be reduced and the wear resistance increased by selecting the right material combinations.

Several researchers [5-7] observed that the friction force and wear rate depend on roughness of the rubbing surfaces, relative motion, type of material, temperature, normal force, stick slip, relative humidity, lubrication and vibration. The parameters that dictate the tribological performance of polymer and its composites also include polymer molecular structure, processing and treatment, properties, viscoelastic behavior, surface texture etc. [8-11]. There have been also a number of investigations exploring the influence of test conditions, contact geometry and environment on the friction and wear behavior of polymers and composites. Watanabe [12], Tanaka [13] and Bahadur and Tabor [14] reported that the tribological behavior of polyamide, high density polyethylene (HDPE) and their composites is greatly affected by normal load, sliding speed and temperature. Pihtili and Tosun [15,16] showed that applied load and sliding speed play significant role on the wear behavior of polymer and composites. They also showed that applied load has more effect on the wear than the speed for composites. Several authors [17-22] observed that the friction coefficient of polymers and its composites rubbing against metals decreases with the increase in load though some other researchers have different views. Stuart [23] and other researchers [24-26] showed that value of friction coefficient increases with the increase in load. Friction coefficient and specific wear rate values for different combinations of polymer and its composite were obtained and compared [27]. For all material combinations, it was observed that the coefficient of friction decreases linearly with the increase in applied pressure values. Unal et al. [28,29] reported that the applied load exerts greater influence on the sliding wear of polymer and its composite than the sliding velocity.

Friction and wear behavior of glass fiber reinforced polyster composite were studied and results showed that in general, friction and wear are strongly influenced by all the test parameters such as applied load, sliding speed, sliding distance and fiber orientations [30]. Moreover, it was found that applied normal load, sliding speed and fiber orientations have more pronounced effect on wear rate than sliding distance. Wang and Li [31] observed that the sliding velocity has more significant effect on the sliding wear as compared to the applied load and variations of wear rate with operating time can be distinguished by three distinct periods. These periods are running-in period, steady-state period and severe wear period, respectively. Tsukizoe and Ohmae [32] showed that reinforcement of fiber or filler significantly improve the tribological behavior of polymeric material but this is not necessarily true for all cases. Suresha et al. [33] showed that there is a strong inter-dependence on the friction coefficient and wear loss with respect to the applied loads for steel-composites contact.

Friction process with vibration is an important practical phenomenon because the influence of vibration can cause significant change in this process. It is known that vibration and friction are dependent on each other. Friction generates vibration in various forms, while vibration affects friction in turns. Some explanations [34-38] are given in order to justify the decrease in the friction coefficient under vibration condition though some of the researchers have different views. Skare and Stahl [39] claimed that mean friction force increases as well as decreases depending on the vibration parameters.

Friction may be increased or decreased depending on the sliding pairs and operating parameters. In this chapter, friction coefficient and wear rate of different types of polymer and composite materials sliding against steel counterface are described. Effects of duration of rubbing, normal load, sliding speed, vertical vibration, horizontal vibration, natural frequency of vibration on friction coefficient are discussed. Some correlations of friction coefficient and wear rate are also incorporated in this chapter.

2. Effect of duration of rubbing on friction coefficient

In sliding contacts, friction coefficient varies with duration of rubbing and these variations are different at different normal loads and sliding velocities. Research works were carried out to investigate the friction coefficient with duration of rubbing for different types of composite and polymer materials. Figure 1 shows the variation of friction coefficient with the duration of rubbing at different normal loads for gear fiber [40]. For normal load 10 N, curve 1 shows that during initial stage of rubbing, friction coefficient is low which remains constant for few minutes then increases very steadily up to a maximum value over a certain duration of rubbing and after that it remains constant for the rest of the experimental time [40].

At initial stage of rubbing, friction force is low due to contact between superficial layer of pin and disc and then, friction coefficient increases due to ploughing effect which causes

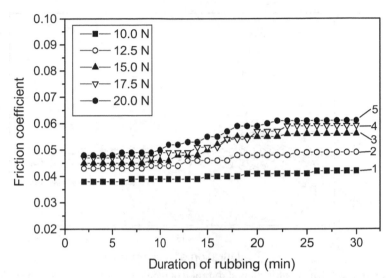

Figure 1. Friction coefficient as a function of duration of rubbing at different normal load, sliding velocity: 1 m/s, relative humidity: 70%, test sample: gear fiber.

roughening of the test disc surface. For normal load 12.5 N, curve 2 shows similar trend as that of curve 1. For normal loads 15, 17.5 and 20 N, curves 3, 4 and 5 show the friction results respectively. The increase in friction coefficient with the increase in normal load is due to the detachment and removal of worn materials and more contact with reinforced cloth fibers and the higher the normal load, time to reach constant friction is less [40]. This is due to the fact that the surface roughness and other parameters attain a steady level at shorter period with the increase in normal load [40].

Figure 2 shows the effect of the duration of rubbing on the value of friction coefficient at different normal loads for glass fiber. For normal load 10 N, curve 1 shows that during initial stage of rubbing, friction coefficient rises for few minutes and then decreases very steadily up to a certain value over some duration of rubbing and then it becomes steady for the rest of the experimental time. Almost similar trends of variation are observed for loads 12.5, 15, 17.5 and 20 N respectively and these results show that friction coefficient decreases with the increase in applied load [40]. It is known that tribological behavior of polymers and polymer composites can be associated with their viscoelastic and temperature-related properties. Sliding contact of two materials results in heat generation at the asperities and hence increases in temperature at the frictional surfaces of the two materials which influences the viscoelastic property in the response of materials stress, adhesion and transferring behaviors [27]. From these results, it can also be seen that time to reach constant friction is different for different normal loads and higher the normal load, glass fiber takes less time to stabilize [40]. Figure 3 shows the variation of friction coefficient with the duration of rubbing at different normal loads for nylon. For 10 N load, curve 1 indicates that during starting of the rubbing, the value of friction coefficient is low which increases for few minutes to a certain value and then decreases almost linearly over some duration of

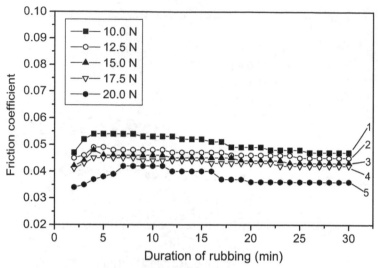

Figure 2. Friction coefficient as a function of duration of rubbing at different normal load, sliding velocity: 1 m/s, relative humidity: 70%, test sample: glass fiber.

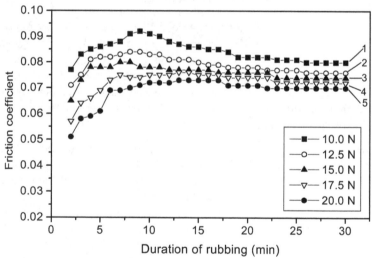

Figure 3. Friction coefficient as a function of duration of rubbing at different normal load, sliding velocity: 1 m/s, relative humidity: 70%, test sample: nylon.

rubbing and after that it remains constant for the rest of the experimental time. Similar trends of variation are observed for normal loads 12.5, 15, 17.5 and 20 N. In these cases, transfer film formed on the stainless steel couterface and the transfer film has important effects on the tribological behavior of a material [22, 40-42]. Friction and wear behavior of polymer sliding against a metal is strongly influenced by its ability to form a transfer film on the counterface [42]. The transfer film formed on a non-polymer counterface is controlled by the counterface material, roughness, and sliding conditions [2].

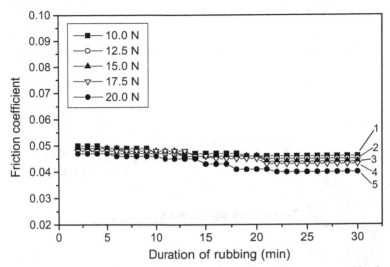

Figure 4. Friction coefficient as a function of duration of rubbing at different normal load, sliding velocity: 1 m/s, relative humidity: 70%, test sample: PTFE.

Figure 4 for PTFE shows that friction coefficient decreases almost linearly up to certain value over some duration of rubbing and after that it remains constant for the rest of the experimental time. It can be noted that transfer film of PTFE formed on the steel counterface due to the strong adhesion across the interface [40,43].

Friction coefficient varies with duration of rubbing at different sliding speeds for different composite and polymer materials [44]. These results are presented in Figs. 5-8.

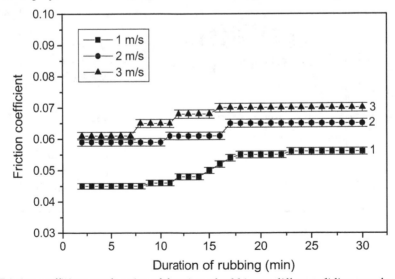

Figure 5. Friction coefficient as a function of duration of rubbing at different sliding speeds, normal load: 15 N, relative humidity: 70%, test sample: gear fiber.

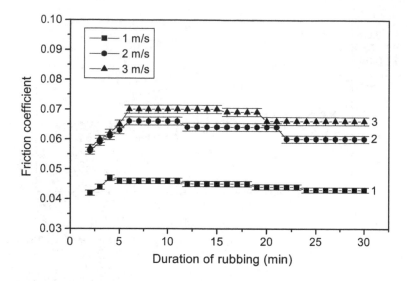

Figure 6. Figure 6. Friction coefficient as a function of duration of rubbing at different sliding speeds, normal load: 15 N, relative humidity: 70%, test sample: glass fiber.

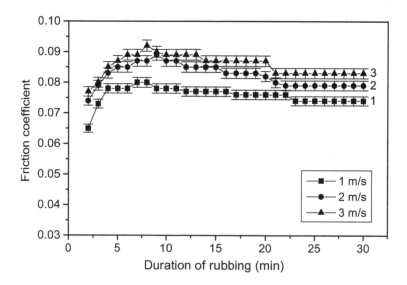

Figure 7. Friction coefficient as a function of duration of rubbing at different sliding speeds, normal load: 15 N, relative humidity: 70%, test sample: nylon.

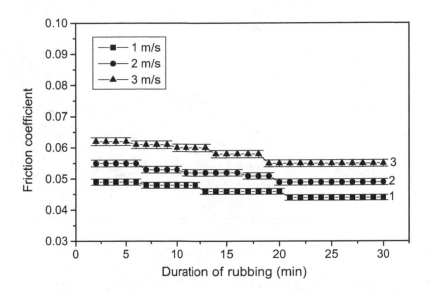

Figure 8. Friction coefficient as a function of duration of rubbing at different sliding speeds, normal load: 15 N, relative humidity: 70%, test sample: PTFE.

3. Effect of normal load on friction coefficient and wear rate

In this section, a comparison of the variation of friction coefficient with normal load for different materials has been discussed. Wear rates of different materials are also compared. Results of Fig. 9 show that friction coefficient decreases with the increase in normal load for glass fiber, PTFE and nylon. Different behavior is observed for gear fiber that is, friction coefficient of gear fiber increases with the increase in normal load. Some factors such as high ploughing, surface damage and breakage of reinforced fibers are responsible for higher friction with higher normal load [40]. Variations of wear rate with normal load for gear fiber, glass fiber, nylon and PTFE are shown in Fig. 10. This figure indicates that wear rate increases with the increase in normal load for all types of materials investigated. The shear force and frictional thrust are increased with the increase in applied load and these increments accelerate the wear rate. Figure 10 also shows the comparison of the variation of wear rate with normal load for gear fiber, glass fiber, nylon and PTFE. The highest values of wear rate for PTFE and lowest values for nylon are obtained among these materials. In case of composite materials, the values of wear rate are higher for gear fiber compared to that of glass fiber. For plastic materials, higher values are observed for PTFE compared to nylon [40].

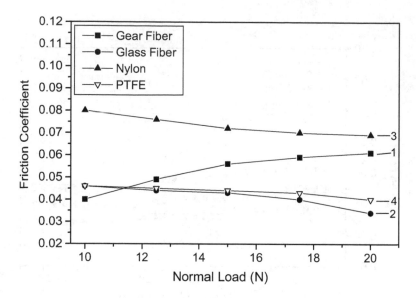

Figure 9. Friction coefficient as a function of normal load for different materials, sliding velocity: 1 m/s, relative humidity: 70%.

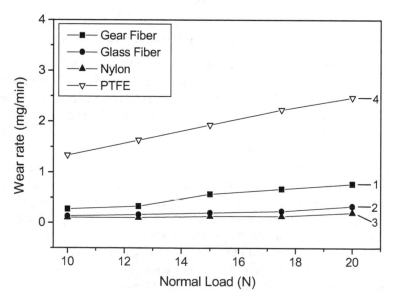

Figure 10. Wear rate as a function of normal load for different materials, sliding velocity: 1 m/s, relative humidity: 70%.

4. Effect of sliding speed on friction coefficient and wear rate

In sliding contacts, sliding speed has an important role on friction and wear of different polymer and composite materials. Figure 11 shows the comparison of the variation of friction coefficient with sliding speed for different materials. Results show that friction coefficient increase almost linearly with sliding speed [44]. These findings are in agreement with the findings of Mimaroglu et al. [27] and Unal et al. [45]. With the increase in sliding speed, the frictional heat may decrease the strength of the materials and high temperature results in stronger or increased adhesion with pin [27,43]. The increase of friction coefficient with sliding speed can be explained by the more adhesion of counterface pin material on disc. Figure 11 indicates that nylon shows the highest friction coefficient within the observed range of sliding speed. Results also reveal that PTFE shows the lowest friction coefficient among these four materials except at sliding speed 1 m/s. At sliding speed 1 m/s, glass fiber shows slightly lower friction coefficient than PTFE but at a sliding speed 3 m/s, glass fiber exhibits much higher friction coefficient than PTFE. This may be due to the breakage of reinforced glass fibers with the increase in sliding speed. Results also show that friction coefficient of gear fiber is higher than that of glass fiber and PTFE. This is due to ploughing effect and breakage of the exposed reinforced cloth fiber of the fracture material [44]. Variations of wear rate with sliding speed for gear fiber, glass fiber, nylon and PTFE are presented in Fig. 12. This figure shows that wear rate increases with the increase in sliding speed for all types of materials investigated. These findings are in agreement with the findings of Mimaroglu et al. [27] and Suresha et al. [33]. The shear force, frictional heat and

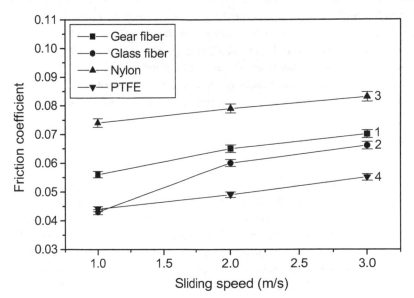

Figure 11. Friction coefficient as a function of sliding speed for different materials, normal load: 15 N, relative humidity: 70%.

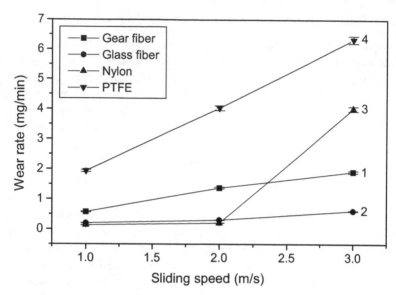

Figure 12. Wear rate as a function of sliding speed for different materials, normal load: 15 N, relative humidity: 70%.

frictional thrust are increased with the increase in sliding speed and these increments accelerate the wear rate. Figure 12 also shows the comparison of the variation of wear rate with sliding speed for gear fiber, glass fiber, nylon and PTFE. From this figure it is observed that PTFE has the highest wear rate among these four materials within the observed range of sliding speed. It is also observed that nylon has the lowest wear rate among these four materials except at sliding speed 3 m/s. At sliding speed 3 m/s, wear rate of nylon is higher than that of gear fiber and glass fiber. Because of higher sliding speed, loss of strength is higher for nylon [43] compared to that of other materials. In case of composite materials, the values of wear rate are higher for gear fiber compared to that of glass fiber [44].

5. Effect of vertical vibration

Figure 13 shows the pin-on-disc set-up with vertical vibration arrangement [46]. For generating vertical vibration, there are two circular plates near the bottom end of the shaft. The upper circular plate fitted with the bottom end of the shaft has a spherical ball in such a way that the ball is extended from the lower surface of this plate. On the top surface of the lower circular plate there are a number of slots. When the shaft rotates, the ball of the upper circular plate slides on the slotted surface of the lower circular plate and due to the spring action, the shaft along with the rotating plate vibrates. The direction of vibration is vertical, i.e. perpendicular to the sliding direction of the pin. By varying the shaft rotation and the number of slots of the lower circular vibration generating plate, the frequency of vibration is varied. By adjusting the height of this slotted plate, the amplitude of the vibration is varied.

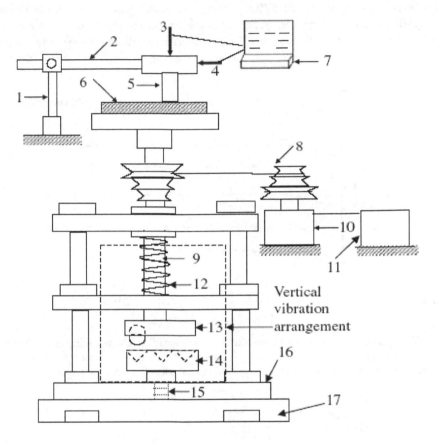

Figure 13. Schematic diagram of the experimental set-up for vertical vibration (1) Load arm holder (2) Load arm (3) Normal load (4) Horizontal load (5) Pin sample (6) Test disc with rotating table (7) Computer (8) Belt and pulley (9) Main shaft (10) Motor (11) Speed control unit (12) Compression spring (13) Upper plate with ball (14) Lower plate with V-slots (15) Height adjustable screw (16) Base plate (17) Rubber block

5.1. Friction coefficient with duration of rubbing at different amplitudes of vibration

The presence of external vertical vibration affects the friction force of different materials considerably which is discussed in this section. Figures 14–17 show the variation of friction coefficient with the duration of rubbing and amplitude of vibration for different types of material. Variations of friction coefficient with time of rubbing for glass fiber reinforced plastic (glass fiber), cloth reinforced ebonite (gear fiber), PTFE and rubber are presented in Figs. 14–17, respectively. These figures indicate that time to reach steady-state value is different for different materials.

Research works carried out for different materials at different frequencies of vibration and these results show that time to reach constant friction is same for these materials [46]. Results also reveal that friction coefficient decrease with the increase in amplitude of vibration. This is due to the fact that the greater the amplitude of vibration, the higher the

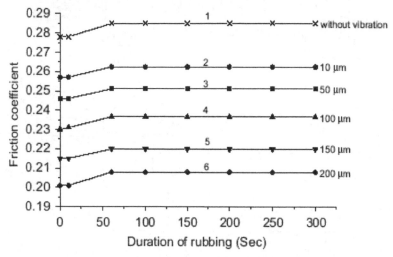

Figure 14. Variation of friction coefficient with the variation of duration of rubbing at different amplitude of vibration (sliding velocity: 0.785 m/s, normal load: 10 N, frequency of vibration: 500 Hz, roughness: 0.2 μm (RMS), relative humidity: 50%, test sample: glass fiber reinforced plastic).

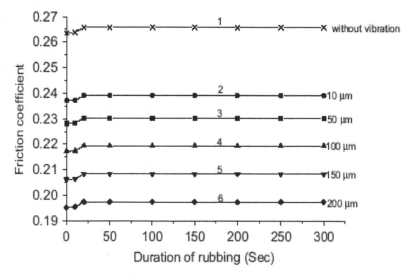

Figure 15. Variation of friction coefficient with the variation of duration of rubbing at different amplitude of vibration (sliding velocity: 0.785 m/s, normal load: 10 N, frequency of vibration: 500 Hz, roughness: 0.2 μm (RMS), relative humidity: 50%, test sample: cloth reinforced ebonite).

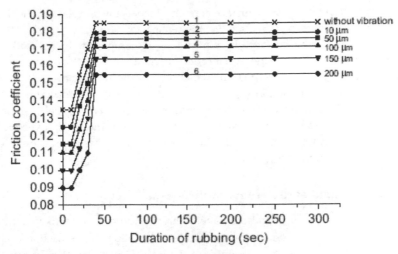

Figure 16. Variation of friction coefficient with the variation of duration of rubbing at different amplitude of vibration (sliding velocity: 0.785 m/s, normal load: 10 N, frequency of vibration: 150 Hz, roughness: 1.5 μm (RMS), relative humidity: 50%, test sample: PTFE).

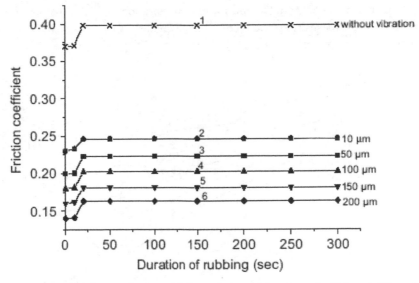

Figure 17. Variation of friction coefficient with the variation of duration of rubbing at different amplitude of vibration (sliding velocity: 0.0393 m/s, normal load: 10 N, frequency of vibration: 50 Hz, roughness: 1.5 μm (RMS), relative humidity: 50%, test sample: rubber).

actual rubbing time, because there is always more separation between the rubbing surfaces due to reduction in the mean contact area of the two sliding objects for vibration [38]. Therefore, the reduction of friction coefficient for the increase in amplitude of vibration is due to the separation of contact surfaces as the higher the amplitude the higher the separation of rubbing surfaces. In fact the higher the separation, the higher the time of contact between the rubbing surfaces is required. As the amplitude increases, keeping the frequency of vibration constant, the acceleration of vibration will also increase that might cause momentary vertical load reduction, which causes the reduction of effective normal force resulting reduction of friction coefficient with the increase of amplitude of vibration. The factors responsible for this momentarily load reduction are: (i) superposition of static and dynamic force generated during vibration, (ii) reversal of the friction vector, (iii) local transformation of vibration energy into heat energy, and (iv) approaching excitation frequency to resonance frequency, etc.

5.2. Friction coefficient at different amplitudes and frequencies of vibration

Amplitude and frequency of vibration have a major role on friction coefficient which is discussed in this section. The effects of amplitude of vibration on the friction coefficient at different frequencies for different materials are shown in Figs. 18–21. Results represent that friction coefficient decreases with the increase in amplitude of vibration at different frequencies of vibration at different rates for different materials. This decrease in friction coefficient is nearly linear for glass fiber, gear fiber and rubber but that for PTFE is nonlinear and its rate is increasing with higher amplitude of vibration [46]. Results also reveal that friction coefficient decreases with the increase in frequency of vibration. These findings are in agreement the findings for mild steel [7].

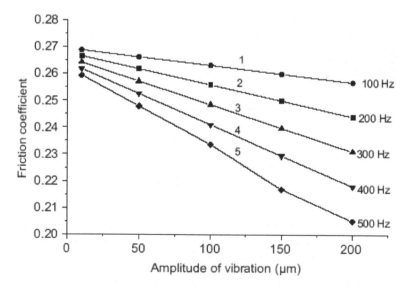

Figure 18. Variation of friction coefficient with the variation of amplitude of vibration at different frequency of vibration (sliding velocity: 1.17 m/s, normal load: 10 N, roughness: 0.2 μm (RMS), relative humidity: 50%, test sample: glass fiber reinforced plastic).

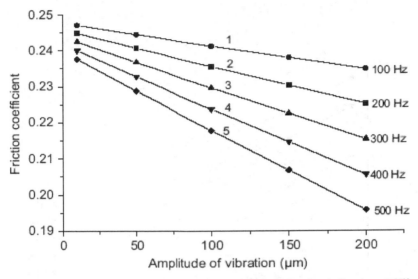

Figure 19. Variation of friction coefficient with the variation of amplitude of vibration at different frequency of vibration (sliding velocity: 1.17 m/s, normal load: 10 N, roughness: 0.2 μm (RMS), relative humidity: 50%, test sample: cloth reinforced ebonite).

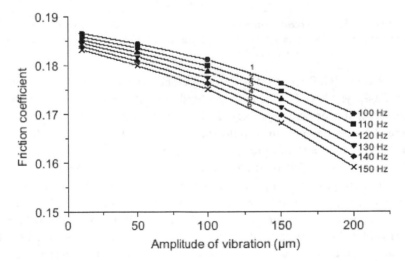

Figure 20. Variation of friction coefficient with the variation of amplitude of vibration at different frequency of vibration (sliding velocity: 1.17 m/s, normal load: 10 N, roughness: 1.5 μm (RMS), relative humidity: 50%, test sample: PTFE).

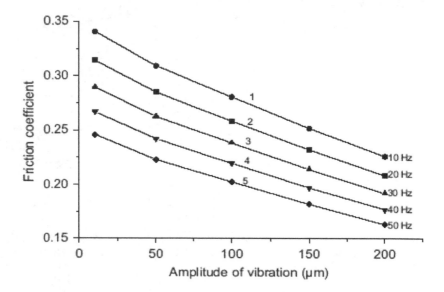

Figure 21. Variation of friction coefficient with the variation of amplitude of vibration at different frequency of vibration (sliding velocity: 0.0393 m/s, normal load: 10 N, roughness: 1.5 μm (RMS), relative humidity: 50%, test sample: rubber).

Friction coefficients of GFRP, mild steel and ebonite are compared for different conditions of vibration (frequency: 100 to 500 Hz and amplitude: 0 to 200 μm) of similar hardness range [47]. These results are presented in Figs. 22-26. Results show that the magnitude and the slope of line of friction coefficient of mild steel are higher than that of GFRP and ebonite. This might be due to the lack of rigidity and strength of the asperities of ebonite and GFRP than mild steel. The variation of friction coefficient with the variation of materials also depends on different physical properties of mating materials and adhesion which depends on inter-atomic force, surface free energy, van der Waals forces, interface condition and chemical interaction due to different types of bonding [43]. It can be noted that at lower frequency (100 Hz), the magnitude of friction coefficient of mild steel is varied significantly than GFRP and ebonite. This variation decreases with the increase in frequency of vibration and at higher frequency (500 Hz), the values friction coefficient of mild steel are almost similar to the friction values of GFRP and ebonite. Under similar conditions, the values of friction coefficient of GFRP are higher than that of ebonite.

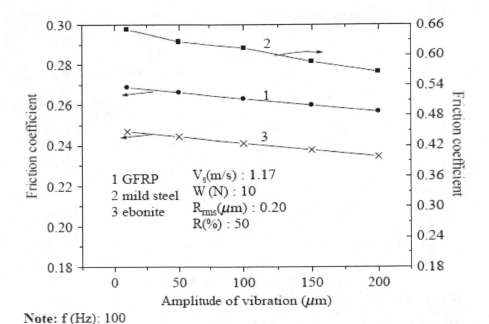

Note: f (Hz): 100

Figure 22. Variation of friction coefficient with the variation of amplitude of vertical vibration

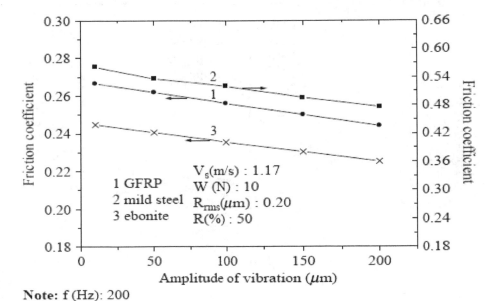

Note: f (Hz): 200

Figure 23. Variation of friction coefficient with the variation of amplitude of vertical vibration

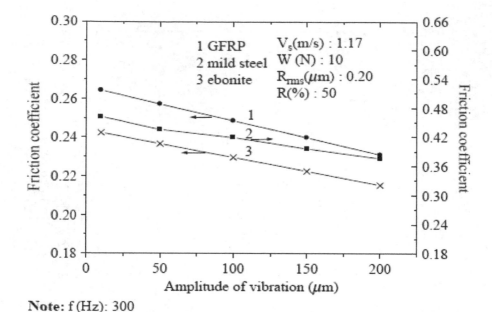

Note: f (Hz): 300

Figure 24. Variation of friction coefficient with the variation of amplitude of vertical vibration

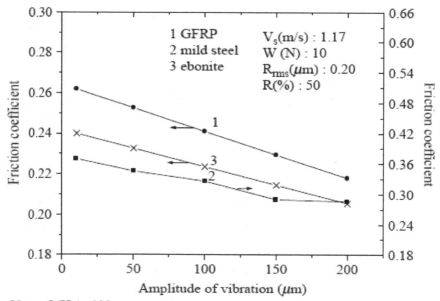

Note: f (Hz): 400

Figure 25. Variation of friction coefficient with the variation of amplitude of vertical vibration

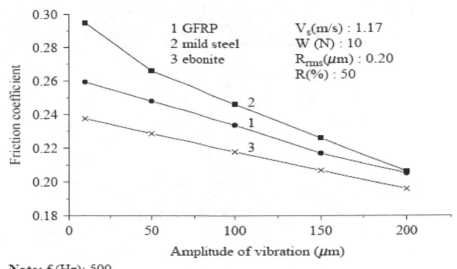

Note: f (Hz): 500

Figure 26. Variation of friction coefficient with the variation of amplitude of vertical vibration

6. Effect of horizontal vibration

The pin-on-disc set-up (Fig. 27) used for generating horizontal vibration [48] in which one end of a coil spring is fixed with the rotating shaft and other end of the spring is fixed with the V-slotted rotating table. An adjusting rigid barrier with spherical tip is fixed with the basic structure of the set-up. Owing to spring action and rotation, the table vibrates horizontally. The direction of vibration is either longitudinal (along the direction of sliding velocity) or transverse (along perpendicular to the direction of sliding velocity) depending on the position of sliding pin on the rotating vibrating table. By varying rotation of the shaft and the number of slots of the rotating table, the frequency of vibration is varied. By adjusting the depth of penetration of the adjustable barrier, the amplitude of the vibration is varied.

The frictional behavior of composite materials under external horizontal vibration is presented in this section. Friction coefficients of GFRP, mild steel and ebonite under longitudinal horizontal vibration are shown in Figs. 28-32. Results show that the friction coefficient increases almost linearly with the increase in amplitude of horizontal vibration for these materials. The increase of friction coefficient might be due to the increase of length of rubbing with the increase of amplitude of vibration. In addition to this the increase of friction coefficient [39,43] is also associated with: (i) Fluctuation of inertia force along the direction of friction force (positive and negative). (ii) More sliding causes more abrasion resistance. Higher abrasion results more shearing due to penetration and ploughing of the asperities between contacting surfaces that might have some effect on the increment of friction force. (iii) Micro-welding, reversal of friction vector, and mechanical interlocking. (iv) Formation and enhance an electrically charge layer at the interface. (v) Increase of solubility due to high temperature.

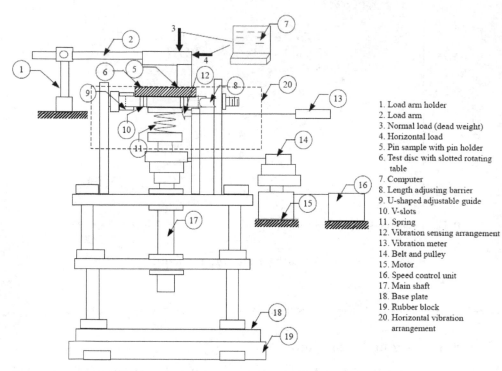

1. Load arm holder
2. Load arm
3. Normal load (dead weight)
4. Horizontal load
5. Pin sample with pin holder
6. Test disc with slotted rotating table
7. Computer
8. Length adjusting barrier
9. U-shaped adjustable guide
10. V-slots
11. Spring
12. Vibration sensing arrangement
13. Vibration meter
14. Belt and pulley
15. Motor
16. Speed control unit
17. Main shaft
18. Base plate
19. Rubber block
20. Horizontal vibration arrangement

Figure 27. Block diagram of the experimental set-up for horizontal vibration

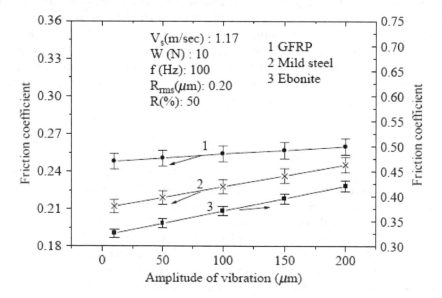

Figure 28. Variation of friction coefficient with the variation of amplitude of longitudinal vibration for frequency 100 Hz

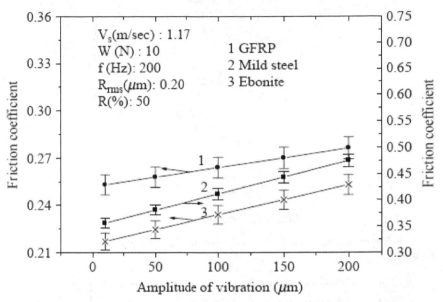

Figure 29. Variation of friction coefficient with the variation of amplitude of longitudinal vibration for frequency 200 Hz

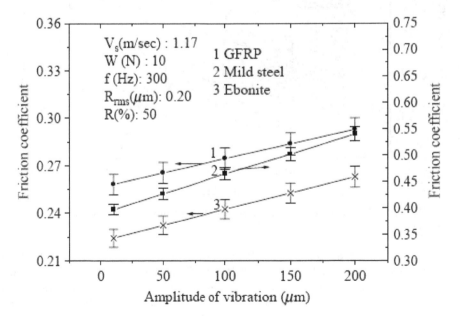

Figure 30. Variation of friction coefficient with the variation of amplitude of longitudinal vibration for frequency 300 Hz

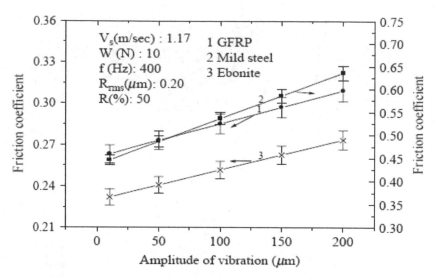

Figure 31. Variation of friction coefficient with the variation of amplitude of longitudinal vibration for frequency 400 Hz

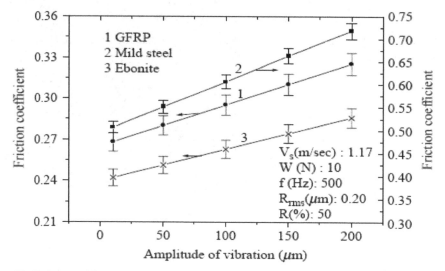

Figure 32. Variation of friction coefficient with the variation of amplitude of longitudinal vibration for frequency 500 Hz

Results indicate that the magnitude and the slope of line of friction coefficient of mild steel under vibration condition are higher than that of GFRP and ebonite. It can be noted that the values of friction coefficient of mild steel are almost twice the values of ebonite and GFRP within the observed range of frequency of horizontal vibration. Under similar vibration conditions, the values of friction coefficient of GFRP are higher than that of ebonite [48]. Results also show that the friction coefficients obtained under transverse vibration are slightly higher than those of longitudinal vibrations. Changing of direction of inertia forces of the vibrating body and the effect of length of sliding path may be responsible for higher friction under transverse vibration [48].

7. Effect of natural frequency on friction coefficient and wear rate

The effects of natural frequency of the experimental set-up on the friction and wear of glass fiber are presented in this section. Figure 33 shows the variation of friction coefficient with the duration of rubbing at different natural frequencies of vibration for glass fiber. Results in Fig. 34 show that friction coefficient increases with the increase in natural frequency of vibration. If a body (either static or dynamic) is in contact with another moving (either rotation or translation) body, where the second body is vibrating, the contact of those two bodies takes place at some particular points of the second body instead of continuous contact. When the natural frequency of vibration of second body is more, for a constant length of contact, the contact points as well as the area of contact between two bodies will be more (Figure 35 (b)) compared to the situation when the natural frequency of second body is less (Figure 35 (a)). As the area of contact or the points of contact between two bodies are more, they experience more frictional resistance for a constant length of contact. Hence, the friction factor between the two bodies will increase with increased natural frequency [49].

The variation of wear and corresponding friction coefficient with the variation of natural frequency of the experimental set-up for glass fiber is presented in Fig. 36. Results show that wear rate as well as friction coefficient increases with the increase in natural frequency of vibration. The shear force and frictional thrust is increased with the increase in natural frequency of vibration and these increased values may accelerate the wear rate. The other possible causes are (i) high ploughing; and (ii) surface damage and breakage of reinforced fibers [50].

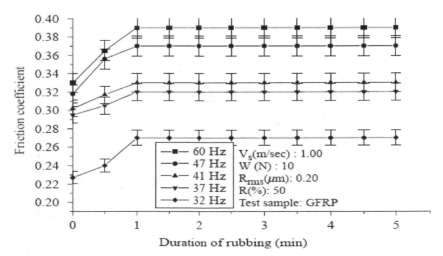

Figure 33. Variation of friction coefficient with the variation of natural frequency of the experimental set-up

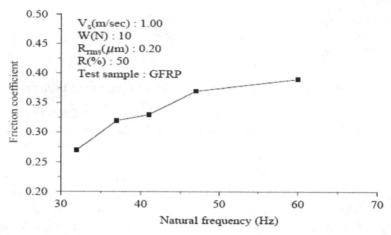

Figure 34. Variation of friction coefficient with the variation of natural frequency of the experimental set-up for GFRP

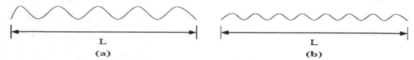

Figure 35. (a) The points of contact of a body with low natural frequency; (b) the points of contact of a body with high natural frequency for constant length of contact (L)

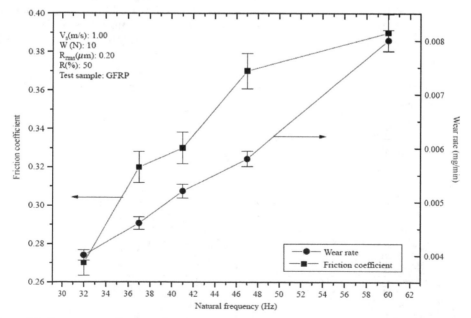

Figure 36. The variation of wear and corresponding friction coefficient with the variation of natural frequency of the experimental set-up for GFRP

8. Dimensionless relationship for friction and wear

8.1. Vertical vibration: Friction coefficient as a function of sliding velocity, amplitude and frequency of vibration

The empirical formula of friction coefficient is derived from the dimensionless analysis to correlate the friction coefficient with sliding velocity, frequency and amplitude of vibration is expressed as [47]:

$$\mu_f = k\left[\frac{Af}{V}\right]^a$$

Where,

μ_f = Friction coefficient
A = Amplitude
V = Sliding velocity
f = Frequency
'a' and 'k' are arbitrary constants

The dimensional friction parameter $\left[\dfrac{Af}{V}\right]$ is called 'Zaman Number' and abbreviated as Zn No.

Figures 37 and 38 show the plot of friction coefficient μ_f versus Zn no. for glass fiber and ebonite, respectively. Figures show that μ_f decreases linearly with the increase of Zn no. and are represented by the equations[47]:

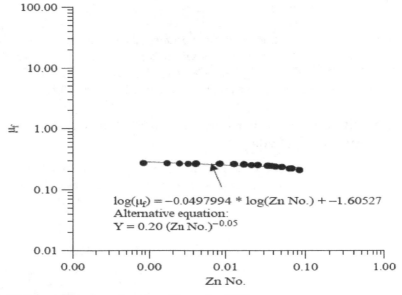

Figure 37. Friction coefficient as a function of Zn no. for GFRP

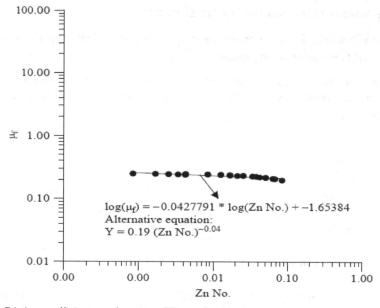

Figure 38. Friction coefficient as a function of Zn no. for ebonite

$$\mu_f = 0.20\left[\frac{Af}{V}\right]^{-0.05} \qquad \text{for glass fiber;}$$

$$\mu_f = 0.19\left[\frac{Af}{V}\right]^{-0.04} \qquad \text{for ebonite}$$

The maximum percentage variation between experimental and theoretical results for GFRP and ebonite are almost ±5% within the observed range of Zn no. The coefficient of determination of GFRP and ebonite are 72% and 75% respectively. This indicates that experimental results are in good agreement with the theoretical calculations.

8.2. Horizontal vibration: Friction coefficient as a function of sliding velocity, amplitude and frequency of vibration

Figures 39 and 40 show the plot of friction coefficient μ_f versus Zn no. for GFRP and ebonite, respectively. Results show that μ_f increases linearly with the increase of Zn no. and are expressed by the equations for external horizontal vibration [48]:

$$\mu_f = 0.34\left[\frac{Af}{V}\right]^{0.05} \qquad \text{for glass fiber;}$$

$$\mu_f = 0.31\left[\frac{Af}{V}\right]^{0.06} \qquad \text{for ebonite}$$

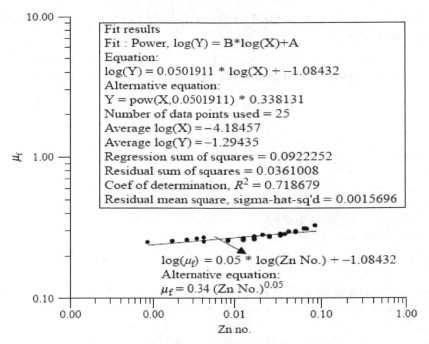

Figure 39. Friction coefficient as a function of Zn no. for GFRP

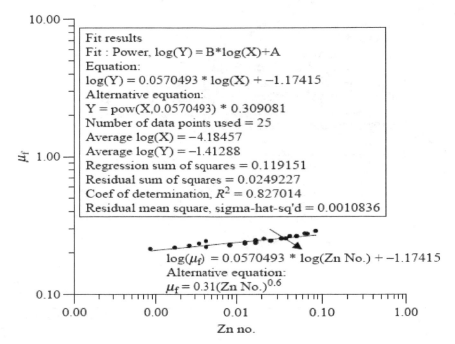

Figure 40. Friction coefficient as a function of Zn no. for ebonite

8.3. Wear rate as a function of natural frequency, sliding velocity, normal load and surface roughness

The empirical formula of wear rate is derived from the dimensional analysis to correlate wear rate with natural frequency, roughness and sliding velocity is expressed as [50]:

$$W_r = \frac{kN}{f_n R}\left[\frac{f_n R}{V}\right]^{-b}$$

Where,
W_r = Wear rate = Mt^{-1}
f_n = Natural frequency = t^{-1}
V = Sliding velocity = Lt^{-1}
N = Normal load = MLt^{-2}
R= Root mean square roughness of the tested surface = L
'b' and 'k' are arbitrary constants

The dimensional wear parameter $\dfrac{f_n R}{V}$ is called 'Asad Number' and abbreviated as Ad No.

Figure 41 shows the plot of wear rate W_r versus Ad No. Results indicate that W_r increases linearly with the increase of Ad. No. and is represented by the equation:

$$Wr = -6.52579E - 4 + 711.092\left[\frac{f_n R}{V}\right]$$

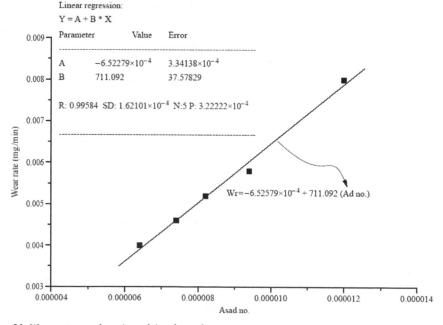

Figure 41. Wear rate as a function of Asad number

The coefficient of determination for the relationship between wear rate and Ad No. is almost 99%. That is, trend line or Ad. No. can explain 99% of the variation in wear rate. This means that experimental results are in good agreement with the theoretical calculations [50].

9. Summary

Friction and wear of polymer and composites are significantly influenced by normal load, sliding velocity, amplitude of vibration, frequency of vibration, direction of vibration and natural frequency. Friction coefficient also depends on duration of rubbing and it is different for different materials. Friction coefficient can be increased or decreased depending on sliding pairs and operating parameters. There are also some correlations between friction/wear and other influencing parameters. The current trends of these experimental and analytical results can be used in future to design different tribological and mechanical components. The researchers can use these results to innovate some design strategies for improving different concerned mechanical processes. It is expected that the research findings of tribological behavior of polymer and composites discussed in this chapter will also be used for future research and development.

Author details

Dewan Muhammad Nuruzzaman* and Mohammad Asaduzzaman Chowdhury
Department of Mechanical Engineering, Dhaka University of Engineering and Technology (DUET), Gazipur, Gazipur, Bangladesh

10. References

[1] Zhang, S. W. State-of-the-art of polymer tribology. Tribology International, 1998; 31 (1–3): 49–60.

[2] Yamaguchi Y., Tribology of plastic materials: their characteristics and applications to sliding components. Amsterdam: Elsevier 1990.

[3] Hooke, C. J., Kukureka, S. N., Liao, P., Rao, M., and Chen, Y. K. The friction and wear of polymers in non-conformal contacts. Wear 1996; 200: 83–94.

[4] Lawrence, C. C., and Stolarski, T. A. Rolling contact wear of polymers: a preliminary study. Wear 1989; 132: 83–91.

[5] J.F. Archard, Wear theory and mechanisms. Wear Control Handbook, ASME, New York, 1980.

[6] D. Tabor, Friction and wear – developments over the last 50 years, keynote address, Proceedings, International Conference of Tribology – Friction, Lubrication and Wear, 50 years on, London, Institute of Mechanical Engineering (1987) 157-172.

[7] Chowdhury MA, Helali MM. The effect of frequency of vibration and humidity on the coefficient of friction. Tribol Int 2006;39:958–62.

* Corresponding Author

[8] N.S.M. El-Tayeb, I.M. Mostafa, The effect of laminate orientations on friction and wear mechanisms of glass reinforced polyester composite. Wear 195 (1996) 186-191.

[9] N.S.M. El-Tayeb, R.M. Gadelrab, Friction and wear properties of E-glass fiber reinforced epoxy composites under different sliding contact conditions. Wear 192 (1996) 112-117.

[10] S. Bahadur, Y. Zheng, Mechanical and tribological behavior of polyester reinforced with short glass fibers. Wear 137 (1990) 251-266.

[11] S. Bahadur, V.K. Polineni, Tribological studies of glass fabric-reinforced polyamide composites filled with CuO and PTFE. Wear 200 (1996) 95-104.

[12] Watanabe, M. The friction and wear Properties of nylon. Wear 1968; 110: 379–388.

[13] Tanaka, K. Transfer of semicrystalline polymers sliding against smooth steel surface. Wear 1982; 75: 183 – 199.

[14] Bahadur, S., and Tabor, D. Role of fillers in the friction and wear behavior of high-density polyethylene. In: Lee LH, editor. Polymer wear and its control. ACS Symposium Series, Washington DC 1985; 287: 253–268.

[15] Pihtili, H., and Tosun, N. Investigation of the wear behavior of a glass fiber-reinforced composite and plain polyester resin. Composites Science and Technology, 2002; 62:367-370.

[16] Pihtili, H., and Tosun, N. Effect of load and speed on the wear behavior of woven glass fabrics and aramid fiber-reinforced composites. Wear, 2002; 252:979-984.

[17] Santner, E., and Czichos, H. Tribology of polymers. Tribology International 1989; 22(2): 103–109.

[18] Tevruz, T. Tribological behaviours of bronze-filled polytetrafluoroethylene dry journal bearings. Wear 1999; 230: 61–69.

[19] Tevruz, T. Tribological behaviours of carbon-filled polytetrafluoroethylene dry journal bearings. Wear 1998; 221: 61–68.

[20] Anderson, J. C. The wear and friction of commercial polymers and composites. In: Friction and wear and polymer composites. Friedrich K, editor. Composite materials series, vol. 1. Amsterdam: Elsevier 1986; 329–362.

[21] Unal, H., Mimaroglu, A., Kadioglu, U., and Ekiz, H. Sliding friction and wear behaviour of polytetrafluoroethylene and its composites under dry conditions. Materials and Design 2004; 25: 239 – 245.

[22] Sirong, Yu, Zhongzhen, Yu, and Yiu-Wing, Mai. Effects of SEBS-g-MA on tribological behavior of nylon 66/organoclay nanocomposites. Tribology International 2007; 40: 855 – 862.

[23] Stuart, B. H. Tribological studies of poly(ether ether ketone) blends. Tribology International 1998; 31(11): 647–651.

[24] Unal, H., and Mimaroglu, A. Influence of test conditions on the tribological properties of polymers. Industrial Lubrication and Tribology 2003; 55(4): 178–183.

[25] Unal, H., and Mimaroglu, A. Friction and wear behavior of unfilled engineering thermoplastics. Material Design 2003; 24: 183–187.

[26] Suresha, B., Chandramohan, G., Prakash, J.N., Balusamy, V., and Sankaranarayanasamy, K. The role of fillers on friction and slide wear characteristics in

glass-epoxy composite systems. Journal of Minerals & Materials Characterization & Engineering 2006; 5 (1): 87 – 101.

[27] Mimaroglu, A., Unal, H., and Arda, T. Friction and wear performance of pure and glass fiber reinforced Poly-Ether-Imide on polymer and steel counterface materials. Wear 2007; 262: 1407 – 1413.

[28] Unal, H., Mimaroglu, A., Kadioglu, U., and Ekiz, H. Sliding friction and wear behaviour of polytetrafluoroethylene and its composites under dry conditions. Materials and Design 2004; 25: 239 – 245.

[29] Unal, H., Sen, U., and Mimaroglu A. An approach to friction and wear properties of polytetrafluoroethylene composite. Materials and Design, 2006; 27: 694-699.

[30] El-Tayeb, N. S. M., Yousif, B. F., and Yap, T. C. Tribological studies of polyester reinforced with CSM 450-R-glass fiber sliding against smooth stainless steel counterface. Wear, 2006; 261:443-452.

[31] Wang, Y.Q., and Li, J. Sliding wear behavior and mechanism of ultra-high molecular weight polyethylene. Materials Science and Engineering, 1999; 266:155–160.

[32] Tsukizoe, T., and Ohmae, N. Friction and wear of advanced composite materials. Fiber Science and Technology, 1983; 18 (4): 265-286.

[33] Suresha, B., Chandramohan, G., Samapthkumaran, P., Seetharamu, S., and Vynatheya, S. Friction and wear characteristics of carbon-epoxy and glass-epoxy woven roving fiber composites. Journal of Reinforced Plastics and Composites 2006; 25: 771-782.

[34] Godfrey, D. (1967), "Vibration reduces metal to metal contact and causes an apparent reduction in friction", ASME Transactions, Vol. 10, pp. 183-92.

[35] Tolstoi, D.M., Borisova, G.A. and Grigorova, S.R. (1973), "Friction reduction by perpendicular oscillations", Soviet Physics-Doklady, Vol. 17, pp. 907-9.

[36] Lenkiewicz, W. (1969), "The sliding process – effect of external vibrations", Wear, Vol. 13, pp. 99-108.

[37] Hess, D.H. and Soom, A. (1991), "Normal vibrations and friction under harmonic loads: Part I – Hertzian contacts", Journal of Tribology, Vol. 113, pp. 80-6.

[38] Budanov, B.V., Kudinov, V.A. and Tolstoi, D.M. (1980), "Interaction of friction and vibration", Soviet Journal of Friction and Wear, Vol. 1, pp. 79-89.

[39] Skare, T. and Stahl, J. (1992), "Static and dynamic friction processes under the influence of external vibrations", Wear, Vol. 154, pp. 177-92.

[40] Nuruzzaman, D.M., Chowdhury, M.A., and Rahaman, M.L. Effect of Duration of Rubbing and Normal Load on Friction Coefficient for Polymer and Composite Materials. Industrial Lubrication. and Tribology 2011; 63: 320 – 326.

[41] Cho M.H., Bahadur, S. and Pogosian A.K. Friction and wear studies using Taguchi method on polyphenylene sulfide filled with a complex mixture of MoS_2, Al_2O_3, and other compounds. Wear 2005; 258: 1825–1835.

[42] Bahadur, S. The development of transfer layers and their role in polymer tribology Wear 2000; 245: 92–99.

[43] Bhushan, B. Principle and Applications of Tribology. John Wiley & Sons, Inc., New York, 1999.

[44] Nuruzzaman, D.M., Rahaman, M.L. and Chowdhury, M.A. Friction coefficient and wear rate of polymer and composite materials at different sliding speeds. International Journal of Surface Science and Engineering 2012 (in press).

[45] Unal, H,. Sen U and. Mimaroglu A. Dry sliding wear characteristics of some industrial polymers against steel counterface Tribology International 2004; 37: 727-732.

[46] Chowdhury M.A.and Helali M.M. The Effect of Amplitude of Vibration on the Coefficient of Friction for Different Materials Tribology International 2008;41: 307- 314.

[47] Chowdhury MA and Helali MM. The frictional behavior of materials under vertical vibration. Industrial Lubrication. and Tribology 2009;61:154–160.

[48] Chowdhury MA and Helali MM. The frictional behavior of composite materials under horizontal vibration. Industrial Lubrication. and Tribology 2009;61:246–253.

[49] Chowdhury MA, Ali M and Helali MM. The effect of natural frequency of the experimental set-up on the friction coefficient Industrial Lubrication. and Tribology 2010;62:78–82.

[50] Chowdhury MA. The effect of natural frequency of the experimental set-up on the wear rate Industrial Lubrication. and Tribology 2010;62:356–360.

Chapter 3

Frequency-Dependent Effective Material Parameters of Composites as a Function of Inclusion Shape

Konstantin N. Rozanov, Marina Y. Koledintseva and Eugene P. Yelsukov

Additional information is available at the end of the chapter

1. Introduction

Electromagnetic composite materials have a number of promising applications in various radio-frequency (RF), microwave and high-speed digital electronic devices, and allow for solving problems related to electromagnetic compatibility (EMC) and electromagnetic immunity (EMI) [1]. For this reason, study and prediction of frequency-dependent radio-frequency RF and microwave properties of materials currently attract much attention. The problem of interest is the analytical description of wideband RF/microwave permittivity and permeability behavior of materials. This is necessary, in particular, to numerically optimize wideband electromagnetic performance of materials and devices at the design stage.

This chapter discusses frequency dependences of effective material parameters (permittivity and permeability) of different types of composites. The chapter consists of three sections. Section I presents a review of approaches for predicting effective material parameters of composites, such as mixing rules, the Bergman–Milton spectral theory, and the percolation theory. Section II suggests on how to select the most appropriate mixing rule for the analysis of properties of a particular composite. Section III considers the dielectric microwave properties of composites filled with fiber-shaped inclusions.

2. Approaches to describe effective material parameters of composites

2.1. Basic mixing rules

In most studies, two-component mixtures are considered, where identical inclusions are imbedded in a homogeneous host matrix. Effective properties of such a composite depend on the intrinsic properties of the inclusions and the host matrix, as well as on the morphology of the composite. The morphology is a characterization of the manner, in which inclusions are distributed in the composite, including their concentration, shape, and correlations in the location. Therefore, the morphology determines how inclusions are shaped and distributed, whether they are mutually aligned/misaligned in the composite, and what concentrations of inclusion phases and a matrix material are.

A conventional approach to describe the properties of composites employs mixing rules, i.e., equations that relates the intrinsic properties of inclusions and the host matrix with the effective properties of composite based on a simple idealized model considering an ellipsoidal-shaped inclusion. Typically, the characterization of the concentration and the shape of inclusions are included explicitly in the mixing rules, and the account for other morphological characteristics is attempted by a proper selection of the mathematical form of mixing rules.

A number of mixing rules are found in the literature. The basic mixing rules are the Maxwell Garnet equation (MG) [2], Bruggeman's Effective Medium Theory (EMT) [3], and the Landau-Lifshitz-Looyenga mixing rule (LLL) [4, 5]. The MG mixing rule,

$$\frac{\chi_{eff}}{1+n\chi_{eff}} = p\frac{\chi_{incl}}{1+n\chi_{incl}},$$
(1)

is equivalent to the Clausius–Mossotti approximation, and also complies with the Ewald-Oseen extinction theorem [6]. Bruggeman's EMT,

$$-(1-p)\frac{-\chi_{eff}}{\chi_{eff}+1-n\chi_{eff}} = p\frac{\chi_{incl}-\chi_{eff}}{\chi_{eff}+1+n\left(\chi_{incl}-\chi_{eff}\right)},$$
(2)

is often referred to as the Polder-van Santen mixing rule in the theory of magnetic composites [7]. The Landau–Lifshitz–Looyenga mixing rule (LLL) is written as

$$\left(\chi_{eff}+1\right)^{1/3}-1 = p\left(\left(\chi_{incl}+1\right)^{1/3}-1\right).$$
(3)

Eqs. (1)–(3) are written for the generalized susceptibilities of inclusions, χ_{incl}, and the effective susceptibility, χ_{eff}, both normalized to the susceptibility of the host matrix, since all susceptibilities of a certain composite, including the effective permittivity and permeability, are governed by the same mixing rule [8], with a possible correction for the tensor nature of the susceptibilities. If permittivity $\varepsilon=\varepsilon'-i\varepsilon''$ is under consideration, then $\chi_{incl}=\varepsilon_{incl}/\varepsilon_{host}-1$ and $\chi_{eff}=\varepsilon_{eff}/\varepsilon_{host}-1$. For the permeability $\mu=\mu'-i\mu''$, $\chi_{incl}=\mu_{incl}-1$ and $\chi_{eff}=\mu_{eff}-1$, because most magnetic composites are based on a non-magnetic host matrix. In Eqs. (1)–(3), n is the form factor, i.e., either depolarization or demagnetization factor, and p is the volume fraction of inclusions.

Starting from the basic mixing rules, simple empirical models of a composite may be suggested.

The MG mixing rule considers the total polarizability of inclusions represented by the right part of Eq.(1) and assumes that this polarizability is acquired to a homogeneous medium. As a consequence, this mixing rule defines the weakest possible cooperative phenomena between neighboring inclusions that are feasible for the given volume fraction of inclusions. The MG mixing rule is an accurate result for the case, when excitation of inhomogeneous fields due to multiple scattering on inclusions and the effect of neighboring inclusions are negligible. Therefore, it coincides with the lower Hashin–Shtrikman limit [9] that provides the smallest possible material parameter in the case, when loss is negligible. The MG mixing rule is believed to be valid for regular composites, i.e., those comprising regularly arranged inclusions, and for the case of conducting inclusions covered with an isolating shell [10, 11].

The physical model for the EMT assumes that the host matrix consists of particles having the same shape that inclusions have, and both the inclusions and the host matrix particles, are embedded in an effective medium with the material constant equal to the effective material constant the composite. The sum of the polarizabilities of these two types of particles must be zero, which corresponds to a homogeneous medium. Though a practical realization of the EMT involves a very special morphology of a composite [12], this mixing rule is widely used, because it incorporates the percolation threshold when modeling a metal-dielectric mixture. The percolation threshold, p_c, is the lowest concentration, at which a macroscopic conductivity appears in the mixture. From the standpoint of the mathematics, the EMT is reduced to a quadratic equation for the effective permittivity. Below and above the percolation threshold, different solutions of the equation must be selected according to the physical selection rules. Equation (2) yields $p_c=n$. In the MG mixing rule (1), the percolation threshold is $p_c=1$.

The LLL mixing rule is built up by an iterative procedure starting from a homogeneous material of inclusions and replacing small amount of this material by the material of the host matrix. After that, the resulting "effective" material is regarded as the homogeneous component for the succeeding substitution step, and so on, which results in Eq. (3). The mixing rule obtained by the same iterative procedure starting with the homogeneous host matrix is referred to as the asymmetric Bruggeman approximation. The result of the LLL mixing rule is independent of the form factor of inclusions. For a metal-dielectric composite, the LLL mixing rule always provides a conductive mixture, so that $p_c=0$.

The LLL mixing rule is known to be an accurate result for the case when the material parameter of inclusions differs slightly from that of the host matrix. In particular, this mixing rule is valid for all material parameters of composites at very high frequencies, because any intrinsic susceptibility of any material approaches zero with the frequency tending to infinity. Agreement of both the MG and EMT mixing rules with the LLL mixing rule in the case of the susceptibility of inclusions slightly differing from zero is attained only if $n=1/3$.

When the volume fraction of inclusions is small, $p \ll p_c$, and the interaction between the inclusions is negligible, all three theories are reduced to the small perturbation limit,

$$\chi_{\text{eff}} = \frac{p\chi_{\text{incl}}}{1+n\chi_{\text{incl}}}.$$

(4)

Strictly speaking, Eqs. (1), (2), and (4) are valid for the case of inclusions of perfectly spherical shape, which have the shape factor equal to 1/3. Otherwise, the equations are not consistent with the LLL mixing rule in the limiting case of high frequencies. For non-spherical particles, the polarizability of inclusions must be averaged over all three principal axes of the inclusion [13]. Two particular cases of non-spherical inclusions are of practical interest - nearly spherical inclusions and highly elongated inclusions (long fibers or platelets). For nearly spherical inclusions, the composites are conventionally described by Eqs. (1), (2), (4) with averaged form factor involved, which is found empirically and may differ from 1/3. For elongated inclusions, the form factor along the shorter axis (in the platelet case), or the sum of two form factors along the shorter axes (in the fiber case) is close to unity, and the polarization of inclusion is these directions can be neglected. In this case, the above equations are valid again, with a randomization factor, κ, included in the right-hand part of the equations to account for an alignment of non-spherical inclusions. For a fiber-filled composite, $\kappa=1/3$, when the fibers are randomly oriented in space, and $\kappa=1/2$, when the fibers are randomly oriented in plane and the wave vector in perpendicular to the plane. For composites filled with platelet-shaped inclusions, $\kappa=2/3$ for the 3D isotropic orientation. For the permeability of composites filled with non-spherical particles, possible anisotropy of magnetic moment, associated with crystallographic anisotropy of particle material, should be also taken into account, see, e.g., [14].

2.2. The Bergman-Milton theory

A generalization of mixing rules may be made with the use of the Bergman-Milton spectral theory (BM) [8]. The theory expresses the effective material parameter of a composite as

$$\chi_{\text{eff}} = p \int_0^1 \frac{\chi_{\text{incl}} b(n) dn}{1 + n \chi_{\text{incl}}}, \tag{5}$$

where the spectral function, $b(n)$, is introduced as a quantitative characterization of the composite morphology. As is seen from Eq. (5), the BM theory accounts for a distribution in effective form factors of inclusions in a composite. This distribution may be associated with the following statistical parameters and processes within the composite: a spread in shapes of individual inclusions comprising the composite; possible agglomeration of inclusions to clusters; and the effects of multiple scattering and inhomogeneous fields excited by neighboring inclusions. Again, the spectral function is the same for all susceptibilities of a particular composite.

The sum rules,

$$\int_0^1 b(n) dn = 1 \quad \text{and} \quad \int_0^1 n b(n) dn = \frac{(1-p)}{D}, \tag{6}$$

relate the spectral function $b(n)$ to the volume fraction of inclusions p for a macroscopically isotropic composite in D dimensions. The practically important cases are $D=3$ (an isotropic 3D composite with non-aligned randomly distributed inclusions, the shape of which is arbitrary in the general case) and $D=2$ (an assembly of infinitely long cylinders). The sum rules provide an agreement of the spectral theory with the LLL mixing rule at $\chi_{\text{incl}} \to 0$.

The spectral theory provides a complete characterization of the frequency dependence of the effective material parameters. The concentration dependence of effective material parameters is implicit in the spectral theory, with the volume fraction involved in the spectral function as a parameter. The analysis of concentration dependences is a powerful tool for understanding properties of composites. However, application of the spectral function approach is not convenient for such analysis.

Another reason that prevents the BM theory from the wide use for the analysis of measured data is that the theory exploits an unknown function, which is difficult to find from the experiment. There are just a few published examples of how to apply the BM theory to the measured data analysis and predicting frequency characteristics of composites [15]. A conventional approach is to accept a functional dependence $b(n)$ as a function of some parameters and to search for these parameters from the measured data [16–18].

Figure 1 shows the calculated spectral functions $b(n)$ for some mixing rules. The spectral function for the MG mixing rule is a delta-function, as is shown in Fig. 1a. The spectral function for the EMT mixing rule presented in Fig. 1b is a semi-circle when plotted as $nb(n)$ against n. Plots d, e, and f show the spectral functions for the McLachlan, Sheng, and Musal–Hahn mixing rules, correspondingly, which are discussed in Subsection 1.4. The latter two plots are composed of several distinct peaks of spectral function even at $n=1/3$. Other examples of calculated spectral functions for mixing rules are found in [19]. In case of elongated inclusions, or if a composite is composed of inclusions with significantly different aspect ratios, the spectral function may consist of two or larger number of separated peaks. Also, several distinct peaks of the spectral function are found to appear due to the interaction between inclusions in periodical composite structures [8].

2.3. The percolation theory

A different approach is provided by the percolation theory, see, e.g., [17]. The percolation theory considers the quasi-static permittivity of a metal-dielectric mixture at concentrations close to the percolation threshold. The main assumption of the theory is that the properties of the material are due to statistical properties of large conductive clusters in this case, rather than due to individual properties of inclusions. The theory predicts a power dependence of static permittivity of the mixture on the difference between the concentration, p, and the percolation threshold, p_c:

$$\varepsilon'_{\text{eff}} \propto \left(p_c - p\right)^{-s}, \quad p < p_c, \quad s \approx 0.7$$
$$\varepsilon''_{\text{eff}} \propto \left(p - p_c\right)^{t}, \quad p > p_c, \quad t \approx 1.8 \tag{7}$$

The values of critical indices, s and t, are believed to be universal, i.e., independent of detailed structure of composite.

A consequence of Eqs. (7) is a power dependence of the real and imaginary permittivity on frequency f:

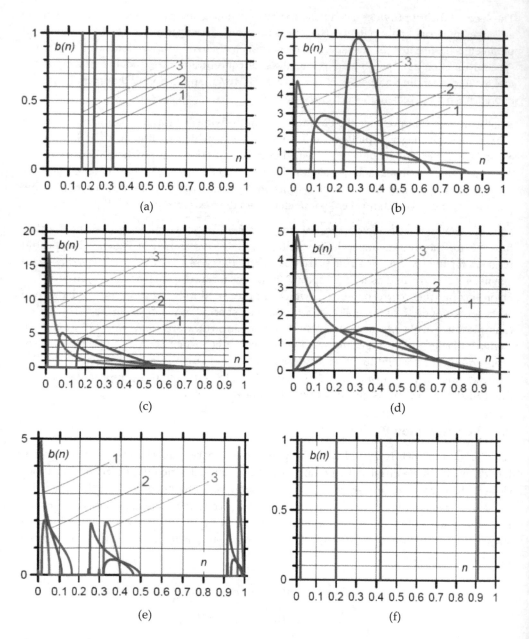

Figure 1. The spectral function, $b(x)$, calculated for various mixing rules: (a) MG, n=1/3: 1 – p=0.1, 2 – p= 0.2, 3 – p=0.5; (b) EMT, n=1/3: p=0.01, 2 – p= 0.1, 3 – p=0.25; (c) Asymmetric Bruggemann's mixing rule: 1 – p=0.1, 2 – p= 0.3, 3 – p=0.6, (d) McLachlan's theory, n=1/3, s=1.8, t=0.7: p=0.01, 2 – p= 0.1, 3 – p=0.25; (e) Sheng theory, n=1/3: 1 – p=0.25, F=0.25,, 2 – p=0.1, F=0.25, 3 – p=0.25, F=0.75; (f) Musal–Hahn theory: n=1/3, p=0.4, F =0.2.

$$\varepsilon'_{\text{eff}} - \varepsilon'_{\text{host}} \propto f^{-Y}, \quad \varepsilon''_{\text{eff}} \propto f^{-Y} \tag{8}$$

The equality of critical indices for the real and imaginary permittivity established by Eq. (8) follows from the Kramers–Krönig relations, if the frequency dependence of permittivity is governed by law (8) within the whole frequency range. In this case, the dielectric loss tangent, $\varepsilon''/\varepsilon'$, is independent of frequency and equal to $\tan(\pi Y/2)$. It follows from the percolation theory that $Y=s/(s+t)\approx0.28$. In practice, observed values of Y are typically closer to zero [20].

The physical reason for the powder dependence of the permeability on frequency may be understood as follows. Assume that the frequency response of an individual inclusion in the composite is governed by the Debye frequency dispersion law,

$$\varepsilon(f) = \varepsilon_\infty + \frac{\varepsilon_0 - \varepsilon_\infty}{1 + if\tau}, \tag{9}$$

where ε_0 is the static permittivity, ε_∞ is the optical permittivity, and τ is the characteristic relaxation time, which is reciprocal to the linear relaxation frequency f_{rel}: $\tau=1/f_{\text{rel}}$. The Debye dispersion law governs the frequency dependence of composites filled with conducting inclusions in most cases. When individual inclusions form large clusters of various sizes, a spread of the characteristic relaxation times τ appears. In this case, the total permittivity is written as:

$$\varepsilon(\omega) = \varepsilon_\infty + \int_0^\infty \frac{B(y)\,dy}{1 + i\omega y}, \tag{10}$$

where $B(y)$ is the distribution function of the relaxation times, and

$$y = \tau/\tau_0 \quad \text{and} \quad \int_0^\infty B(y)\,dy = 1. \tag{11}$$

The cumulative dispersion curve becomes more gently sloping. With a special form of the distribution,

$$B(y) = \frac{y}{2\pi} \frac{\sin \beta\pi}{\cosh(1-\beta)y - \cos \beta\pi}, \tag{12}$$

the Cole-Davidson frequency dispersion, see, e.g., [21], is obtained,

$$\varepsilon(f) = \varepsilon_\infty + \frac{\varepsilon_0 - \varepsilon_\infty}{1 + (if\tau)^{1-\alpha}}. \tag{13}$$

Dispersion law (13) involves a frequency region governed by a power frequency dependence of the permittivity. The form of the distribution does not significantly affect the result provided that the distribution is wide, which can be a kind of justification for the percolation theory.

When the property under consideration is the permeability, the percolation behavior is not readily observed [22].

2.4. Complex mixing rules

There are many practical scenarios when none of the simple mixing laws agree with the measured data on a practical composite. A classical example is related to carbonyl iron composites. Despite almost perfect spherical shape of carbonyl iron particles, the form factor restored from the volume fraction dependence of the permittivity or permeability frequently differs greatly from 1/3 [18, 23]. The reason is an agglomeration of the inclusions.

Another example is the percolation threshold study in composites composed of the same carbon black and different polymer host matrices [24]. Depending on a polymer, the percolation threshold may vary from 5 to 50%. Polymerization with different polymers results in different morphology of the composites. The reason is agglomeration or de-agglomeration of inclusions, which depends on the properties of the interface between inclusions and the host matrix. The importance of spatial distribution of inclusions in a composite for validity of mixing rules is discussed in [25].

Practically, in describing properties of composites, many other factors must be accounted for. Among these factors, there are the distribution of inclusions in shape [26–28] and size [29]; the presence of an oxide layer on the surface of conducting particles, statistical spread of intrinsic material parameters of inclusions, e.g., their conductivities [30], as well as possible cones of orientations, if elongated particles are aligned or randomly oriented [31]. For these reasons, fitting parameters are typically unavoidable in accurate description of material properties of composites.

Therefore, taking into account peculiarities of a composite morphology may be crucial for accurate description of composite performance, especially in the case of permittivity of metal-dielectric mixtures, where the intrinsic permittivity of inclusions is infinity in the quasi-static case, and the effective permittivity of composite is determined solely by the shape of inclusions.

Conventionally, a composite morphology is accounted for using more complex mixing rules, which involve some fitting parameters. Three examples of such mixing rules are discussed below. These theories combine the above mentioned basic approaches, and allow for introducing appropriate fitting parameters.

A well known example of such combination is the Lichtenecker mixing rule [32], which is written for the case of the effective permittivity as

$$\varepsilon_{\text{eff}}^k = p\varepsilon_{\text{incl}}^k + (1-p)\varepsilon_{\text{host}}^k. \tag{14}$$

In Eq. (14), k has a physical meaning of a critical exponent, which is conventionally treated as a fitting parameter to obtain an agreement with measurements. Equation (14) may be considered as an empirical combination of the LLL mixing rule and the percolation theory.

A combination of the EMT mixing rule and the percolation theory is MsLachlan's Generalized Effective Medium Theory [33]:

$$p\frac{\varepsilon_{incl}^{1/t} - \varepsilon_{eff}^{1/t}}{\varepsilon_{eff}^{1/t} + n\left(\varepsilon_{incl}^{1/t} - \varepsilon_{eff}^{1/t}\right)} + (1-p)\frac{\varepsilon_{host}^{1/s} - \varepsilon_{eff}^{1/s}}{\varepsilon_{eff}^{1/s} + n\left(\varepsilon_{host}^{1/s} - \varepsilon_{eff}^{1/s}\right)} = 0 .$$ (15)

In this equation, which is also written for the case of permittivity, the EMT equation (2) is supplemented with percolation indices s and t. These indices, together with the effective depolarization factor of inclusions n, are also treated as the fitting parameters.

Another approach for developing complex mixing rules is to divide inclusions in composite into two groups (e.g., a part of inclusions are considered as isolated and the other part are assumed to compose dense clusters [34], or any other way of subdivision into groups), and then to mix these groups with different mixing rules. The value of F, $0<F<1$, a fraction of inclusions attributed to one of the groups, provides a fitting parameter. An example of this approach is Sheng's theory [35]:

$$FP_1 + (1-F)P_2 = 0,$$ (16)

where P_1 and P_2 are the effective polarizabilities of the two groups of particles; for the case of spherical inclusions, P_1 and P_2 are found in Sheng's theory as

$$P_1 = \frac{\left(\varepsilon_{eff} - \varepsilon_{host}\right)\left(\varepsilon_{incl} + 2\varepsilon_{host}\right) + \left(\varepsilon_{host} - \varepsilon_{incl}\right)\left(\varepsilon_{eff} + 2\varepsilon_{host}\right)p}{\left(2\varepsilon_{eff} + \varepsilon_{host}\right)\left(\varepsilon_{incl} + 2\varepsilon_{host}\right) + 2\left(\varepsilon_{eff} - \varepsilon_{host}\right)\left(\varepsilon_{host} - \varepsilon_{incl}\right)p} ,$$ (17)

$$P_2 = \frac{\left(\varepsilon_{eff} - \varepsilon_{incl}\right)\left(2\varepsilon_{incl} + \varepsilon_{host}\right) + \left(\varepsilon_{incl} - \varepsilon_{host}\right)\left(\varepsilon_{eff} + 2\varepsilon_{host}\right)\left(1-p\right)}{\left(2\varepsilon_{eff} + \varepsilon_{incl}\right)\left(2\varepsilon_{incl} + \varepsilon_{host}\right) + 2\left(\varepsilon_{eff} - \varepsilon_{incl}\right)\left(\varepsilon_{incl} - \varepsilon_{host}\right)\left(1-p\right)} .$$ (18)

Equation (17) describes the polarizability of a spherical particle consisting of an inclusion material and then coated by a shell of the host matrix material. Equation (18) represents the inverse structure, with the shell made of the inclusion material and the core made of the host matrix material. Both Eqs. (17) and (18) are consistent with the MG formalism. The effective structures are mixed with each other according to the EMT equation (16). An analogous approach is suggested by Musal and Hahn [36], with the only difference that the EMT equation (16) describing a mixture of the two groups is substituted in the MG equation (1). Doyle and Jacobs [34, 37] suggested the model, where the two groups of inclusions comprise isolated inclusions and clusters of closely packed inclusions.

The complex mixing rules are suggested and provide rather good/reasonable agreement with measured data mostly for the concentration dependences of the permittivity in metal-dielectric mixtures. However, these theories may fail when describing frequency dependences of material parameters. The reason is that the complex mixing rules have the spectral function consisting of several isolated peaks even in the case of nearly-spherical inclusions, as is seen in Fig. 1 *e* and *f*. A physical meaning can hardly be attributed to these peaks in case of a random composite filled with spherical inclusions, because, as is shown below, the appearance of isolated peaks of spectral function generally results in the appearance of several isolated regions of frequency dispersion of material parameters.

Another approach to the problem of the permittivity dependence on concentration for metal-dielectric mixtures has been suggested by Odelevskiy [38]. He was the first who noticed the analogy between the MG and EMT equations, in which the concentration dependence of the permittivity for conducting inclusions are written as

$$\varepsilon_{\text{eff}} = 1 + \frac{1}{n}\frac{p}{1-p} \tag{19}$$

and

$$\varepsilon_{\text{eff}} = 1 + \frac{1}{n}\frac{p}{1-p/n}, \tag{20}$$

respectively. Odelevskiy suggested an equation that generalizes these two theories in the case of a metal-dielectric mixture:

$$\varepsilon_{\text{eff}} = 1 + \frac{1}{n}\frac{p}{1-p/p_c}. \tag{21}$$

In Eq. (21), the form factor n and percolation threshold p_c are the two fitting parameters. With these fitting parameters, the equation demonstrates an excellent agreement with measured data for a variety of different metal-dielectric mixtures [39], if the concentration of inclusions is not very close to the percolation threshold. Equation (21) cannot be considered as an independent mixing rule, because it does not leave a room for the permittivity of inclusions different from infinity.

3. Frequency-dependent behavior of composites and validity of mixing rules

Effective properties of composites in the majority of mixing rules and theories are considered in the quasi-static approximation. Because of this, the frequency dependence of effective material parameter appears due to the difference in frequency dependences of material parameters of constituents.

Frequency dispersion of permittivity in a composite frequently appears due to the different frequency behavior of its dielectric host matrix and of conducting inclusions. Host matrices are typically considered as non-dispersive over a frequency range of interest, while the permittivity of metallic inclusions is imaginary and reciprocal to frequency. There are other dielectric materials possessing dielectric dispersion at microwaves, e.g., water, some ferroelectrics [40], some lossy polymers, but necessity of accounting for this dispersion is a fairly rare.

In contrast, a multitude of magnetic materials exhibit frequency dispersion of permeability at microwaves. The reason is that all magnets lose their magnetic properties at frequencies below several gigahertz, as is shown in Subsection 2.2. These are the microwaves, or even lower frequencies, where the permeability changes from large static permeability to unity. Notice that the intrinsic permeability of magnetic powders is generally unknown. It depends not only on the composition of the material, but also on manufacturing and treatment technology, and the latter dependence may be essential.

In the first-order approximation, the frequency dependence of material parameters may be considered as an assembly of loss peaks accompanied by corresponding frequency dispersion of the real part, according to the Kramers-Kronig relations. In many cases, the Lorentzian (resonance) dispersion law,

$$\chi(f) = \sum_{i=1}^{m} \frac{\chi_{\mathrm{st},i}}{1 + i f/f_{\mathrm{rel},i} - \left(f/f_{\mathrm{res},i}\right)^2}, \tag{22}$$

provides a good fitting of measured dependences of susceptibility χ on frequency f. In Eq. (22), m is the number of the resonance terms involved in the dispersion law, and $\chi_{\mathrm{st},i}$, $f_{\mathrm{rel},i}$, and $f_{\mathrm{res},i}$ are the static susceptibility, relaxation frequency, and resonance frequency attributed to i-th resonance term, respectively.

3.1. Frequency-dependent behavior of composites

Almost all mixing rules deduce the effective material parameters from the polarizability,

$$P = \frac{\chi_{\mathrm{incl}}}{1 + n\chi_{\mathrm{incl}}}, \tag{23}$$

embedded in either a host matrix or the effective medium. From Eq. (23), two limiting cases are clearly seen, $n\chi_{\mathrm{incl}} \ll 1$ and $n\chi_{\mathrm{incl}} \gg 1$.

In the case of $n\chi_{\mathrm{incl}} \ll 1$, the LLL mixing rule (3) is a rigorous result. For majority of practical cases, Eq. (3) may be rewritten just as the perturbation limit given by Eq. (4). In this case, the effective material parameter is just the intrinsic material parameter multiplied by the volume fraction of inclusions. This means that the effect of interaction between inclusions is negligible. The morphology of the composite, including the shape of inclusions, does not affect the effective material parameter. This case is typical for the microwave permeability of composites filled with either fibrous or platelet inclusions, as well as for all effective susceptibilities at very high frequencies.

In the other limiting case, $n\chi_{\mathrm{incl}} \gg 1$, the effective material parameter depends on the morphology only. Here, the effective static susceptibility increases non-linearly with the concentration of inclusions, according to the percolation behavior and Odelevskiy equation (21). It is the case, for which most of the complex mixing rules have been developed. The case is related to the permittivity of metal-dielectric mixtures, since the imaginary part of the permittivity of metal inclusions is so high that the absolute value of the microwave permittivity can be considered as infinite. As to the permeability, this case may be observed in some composites filled with ferromagnetic inclusions of spherical shape, or for low-frequency magnetic materials, whose permeability may be very high.

Let the effective material parameter be considered in a wide frequency range. Assume that the host matrix of the composite is lossless and non-dispersive. Then, the frequency dispersion in the composite is due to the frequency dispersion of inclusions. It is well known that a material parameter of any medium approaches unity with the frequency tending to infinity. Because of that, the case of $n\chi_{incl}\ll 1$ is always observed at very high frequencies, where the LLL mixing rule (3) describes material parameters of composites.

If the intrinsic susceptibility of inclusions is low, the LLL mixing rule is valid for low frequencies as well. In this case, the frequency dependence of any effective material parameter is just proportional to the dependence for the intrinsic material parameter, and the volume fraction of inclusions is the coefficient of proportionality. The loss peak in the composite and the loss peak of inclusions are located at the same frequency. The concentration dependence of the effective parameter is linear over the entire frequency range.

Another possibility is when the inequality $n\chi_{incl}\gg 1$ holds at low frequencies. In this case, the frequency dispersion in the composite appears, when the absolute value of $n\chi_{incl}$ is about unity. The loss peak in the composite is shifted towards higher frequencies as compared to the loss peak of inclusions. As the concentration of inclusions increases, the loss peak is shifted to the lower frequencies. At frequencies above the peak, the effective susceptibility is again proportional to the intrinsic susceptibility. At frequencies below the peak, the effective permeability depends mostly on the composite morphology and is independent of the intrinsic susceptibility. The concentration dependence of the effective susceptibility is non-linear.

This case is typical for metal-dielectric mixtures. However, the conductivity of metals is usually too high to provide a loss peak of permittivity at microwaves. The microwave permittivity for most metal-dielectric composites may be considered as non-dispersive and low-loss. An exception is the percolation behavior, which will be discussed in Section 3.

If the frequency dependence of the intrinsic material parameter is Lorentzian (22) with $m=1$, and the mixing rule describing the composite is the MG, then the frequency dependence of the effective material parameter is Lorentzian as well. The parameters of the dispersion law for the effective material parameters are given by the simple equations [41]

$$\chi_{st,eff} = \frac{\chi_{st,incl}p}{\chi_{st,incl}\, n(1-p)+1}, \tag{24}$$

$$f_{rel,eff} = f_{rel,incl}\left(\chi_{st,incl}\, n(1-p)+1\right), \tag{25}$$

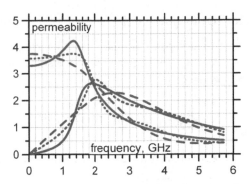

Figure 2. The frequency dependence of effective permeability of a composite calculated by the EMT (2) with p=0.25 and n=1/3 (solid curves, blue curve for real permeability and red curve for imaginary permeability). Inclusions in the composite exhibit the Lorentzian frequency dispersion (22) with m=1, χ_{st}=100, f_{res}=1 GHz, and f_{rel}=2 GHz. The dashed curves are the best fit of the solid line with the Lorentzian dispersion law with m=1, the dotted curves – with a sum of two Lorentzian terms, m=2.

and

$$f_{res,eff} = f_{res,incl} \sqrt{\chi_{st,incl}\, n(1-p)+1}\,, \tag{26}$$

where subscript "incl" indicates the Lorentzian parameters of the intrinsic permeability of inclusions and subscript "eff" is related to the Lorentzian parameters of the effective susceptibility of composite. It is clearly seen from the equations that the non-linear concentration dependence of static susceptibility is accompanied by a low-frequency shift of both the characteristic frequencies. A general validation of this fact is given in the next subsection.

As is seen from Eqs. (25) and (26), the MG mixing rule retains the shape of effective susceptibility loss peak characteristic for the intrinsic susceptibility of inclusions. From the standpoint of the BM spectral theory, the reason is that additional loss due to mixing may arise over the entire range of effective form factors, where the spectral function has non-zero values. The spectral function for the MG mixing law is a delta-function, therefore, additional loss, which may distort the loss peak, does not appear.

Other mixing rules are characterized by a spectral function of a finite width and may therefore result in distorted shape of the loss peak. Figure 2 shows the frequency dependence of effective permeability of a composite calculated by the EMT (2). Inclusions in the composite are assumed to exhibit the Lorentzian frequency dispersion (22) with m=1. In the figure, the dashed curves are the best fit of the calculated permeability with the Lorentzian dispersion law with m=1, the dotted curves are obtained for the sum of two Lorentzian terms, m=2. It is seen that the EMT produces a large distortion of the Lorentzian dispersion curve, when the concentration is close to the percolation threshold. The distortion has a form of the increased loss at the high-frequency slope of the loss peak, because the spectral function peak for the EMT is extended to the region of large arguments, see Fig. 1b.

3.2. Integral relations for the frequency dependences in composites

The low-frequency shift of the loss peak appearing with increasing volume fraction and accompanied by non-linear concentration dependence of static susceptibility is a general rule. Let us consider two integrals,

$$I_1 = \frac{2}{\pi}\int_0^\infty \chi'' f\,df \quad \text{and} \quad I_2 = \frac{2}{\pi}\int_0^\infty \chi'\,df, \tag{27}$$

which are analogous to the well-known sum rule for the Kramers-Kronig relations,

$$\chi'_{st} = \frac{2}{\pi}\int_0^\infty \frac{\chi''df}{f}. \tag{28}$$

The difference between (28) and (27) is that the values of I_1 and I_2 are determined by the high-frequency asymptote of the susceptibility, rather than by the low-frequency asymptote, which defines the value of integral (28). In composites, this asymptotic behavior is governed by the LLL mixing law. Therefore, integrals I_1 and I_2 for any composite are equal to the corresponding values for the bulk material of inclusions multiplied by the volume fraction of inclusions [42,43]

$$I_{i,composite} = pI_{i,inclusions}. \tag{29}$$

Consideration of Eq. (29) makes sense if the integrals are convergent and have a non-zero value. For I_1, this is true for the Lorentzian dispersion law (22) that has the high-frequency asymptote given by:

$$\chi(f) \approx -\chi_{st}\left(\frac{f_{res}}{f}\right)^2 + i\chi_{st}\left(\frac{f_{res}}{f}\right)^3 \frac{f_{res}}{f_{rel}}. \tag{30}$$

For I_2, the convergence is provided by the Debye dispersion law (9), which is the limiting case of (22) at $f_{res}\to\infty$ and has the high-frequency asymptote represented as:

$$\chi(f) \approx i\chi_{st}\left(\frac{f_{rel}}{f}\right) + \chi_{st}\left(\frac{f_{rel}}{f}\right)^2. \tag{31}$$

In the theory of magnetic material, these integrals are employed to validate ultimate values of high-frequency permeability. The corresponding constants for magnetic materials depend on the saturation magnetization of the material, M_s. If the frequency dependence of effective permeability is either single-term Lorentzian or Debye, then the values of the integrals are related to the static magnetic susceptibility and the resonance frequency

$$I_1 = p\kappa(\gamma M_s)^2 \approx \chi_{st,eff}\,f_{res,eff}^2, \tag{32}$$

$$I_2 = p\kappa\left(\gamma M_s\right) \approx \chi_{\text{st,eff}} f_{\text{rel,eff}}. \tag{33}$$

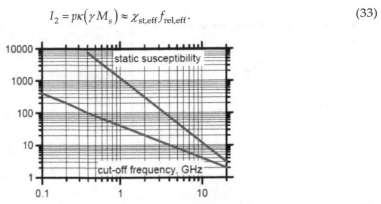

Figure 3. The static susceptibility as a function of cut-off frequency calculated with Acher's law (32) (red line, κ=1/3) and Snoek's law (33) (blue line, κ=2/3). In both the cases, M_s=2.15 T, p=1.

In Eqs. (32) and (33), $\gamma\approx3$ GHz/kOe is the gyromagnetic ratio, and κ is the randomization factor. For I_2, typically κ=2/3; for I_1, different possibilities are discussed in [14]. Equations (32) and (33) represent the well known Acher's law [44] and Snoek's law [45], respectively. Then I_1 has a meaning of Acher's constant, and I_2 is Snoek's constant. For most materials, Snoek's law is valid, which involved Debye frequency dependence and integral I_2. For some materials, such as hexagonal ferrites and thin ferromagnetic films, Acher's law is valid, so that integral I_1 is calculated as Eq. (32), and much larger high-frequency permeability values can be obtained. The laws (32) and (33) are used for estimating high-frequency magnetic behavior of materials. A magnetic material may have high permeability value at frequencies below the cut-off frequency, which is the least of f_{res} and f_{rel}, where the permittivity falls to values close to unity. As the saturation magnetization of magnetic materials is typically below approximately 2 T, it follows from (32) and (33) that magnetic materials with high static permeability are permeable at frequencies of microwave range or lower.

Figure 3 shows the ultimate values of the static magnetic susceptibility as a function of the cut-off frequency calculated with Acher's law (32), red line, and Snoek's law (33), blue line. In both cases, M_s=2.15 T and p=1, which corresponds to a homogeneous sample of pure iron. For Snoek's law, κ=2/3; for Acher's law, κ=1/3 is accepted, which corresponds to random distribution of thin platelets. It is seen from the figure that with low values of the cut-off frequency, below 1 GHz, Acher's law enables a large advantage over Snoek's law in feasible values of the static permeability. At higher frequencies, this advantage eliminates, and both the laws permits rather small ultimate values of static permeability with cut-off frequencies of several dozen gigahertz.

For the permittivity of a metal-dielectric mixture, the frequency dependence is of Debye type, and an analogue of Snoek's law may be introduced. As $\varepsilon''_{\text{incl}}=2\sigma/f$, where σ is the conductivity of inclusions, the analogue of Snoek's constant for permittivity would be just the doubled conductivity of inclusions,

$$I_2 = 2p\sigma. \tag{34}$$

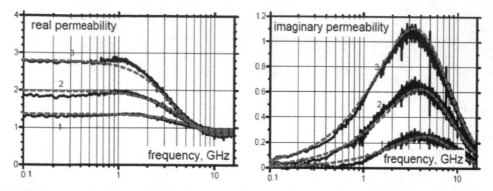

Figure 4. The measured frequency dependence of permeability of hexagonal ferrite composites (black curves, left: the real part, right: the imaginary part) for the Co_2Z composites. The volume fractions of ferrite are: 1, p=0.1; 2, p=0.3; and 3, p=0.5. The red curves show the results of fitting the measured data with the Lorentzian dispersion law (22) with m=1 [41].

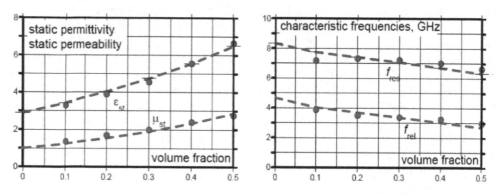

Figure 5. Left: the static permittivity (red dots) and static permeability (blue dots); right: the resonance frequency (red dots) and relaxation frequency (blue dots). The data are obtained for hexagonal ferrite composites by fitting the measured frequency dependences of permeability with the Lorentzian dispersion law (22) with m=1. The curves are the best fit of corresponding dots with Eqs. (24–26) [41].

3.3. Applicability of the MG mixing rule

The MG mixing rule usually agrees closely with the measured data, when $n\chi_{incl}$ ~1. This is a frequent occasion for the microwave permeability of magnetic composites. The intrinsic permeability of magnetic materials does not exceed several units at microwaves due to the fast decrease with frequency, according to Snoek's and Acher's laws. With these relatively low intrinsic permeability values, the dependence of the effective material parameters on the shape of inclusions appears. In particular, a low-frequency shift of the loss peak is observed as p increases. However, the dependence is still weak and may therefore be characterized by an averaged demagnetization factor n.

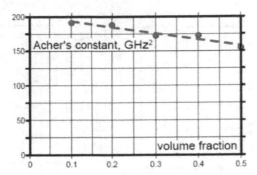

Figure 6. The measured ratio of Acher's constant to the volume fraction plotted against the volume fraction (dots). The line show the linear fit of the measured data [41].

An example of measured data having a good agreement with the MG mixing rule is taken from [41], where composites filled with powders of hexagonal ferrite have been studied. Figure 4 shows the measured microwave permeability for three of the samples. Application of the single-term Lorentzian dispersion law (22) provides a good agreement with the measured data for all volume fractions. This is seen from Fig. 5, where the static permittivity and the Lorentzian characteristic frequencies, obtained by the best fits of the measured magnetic dispersion curves, are plotted as functions of volume fraction. The curves in the figure are obtained by fitting the experimental points (dots) with Eqs. (24–26). For the bulk hexagonal ferrite, the retrieved static values are $\varepsilon_{st,incl}=16$ and $\mu_{st,incl}=11$, and $n\approx0.33$, which indicates that the ferrite particles are of nearly spherical shape. Therefore, $n\chi_{st,incl}$ is the range from 3 to 5, and is reasonably close to unity. The measured data on the microwave material parameters of the composites under study agree with the MG mixing rule calculations, which is evidenced by close agreement of the dots and the fitting curves in Fig. 5.

However, an accurate analysis of the data reveals some disagreement. Acher's constant of the composites, calculated from the data for different volume fractions does not agree with Eq. (32). As Fig. 6 shows, Acher's constant depends on the volume fraction of inclusions, which should not be the case. The reason could be a distribution of shapes of individual inclusions that may result in deviation of the morphology from that postulated in the MG approach. This problem is discussed in more details in the next Subsection.

3.4. Account for the distribution in shapes of inclusions

A case, which may require a sophisticated mixing rule, is a composite filled with conducting ferromagnetic inclusions, whose both permittivity and permeability must be predicted, for example, to describe electromagnetic performance of the composite. In this case, two products of the form factor and the static susceptibility are involved, for the dielectric and magnetic susceptibility, which enlarges the range of variation of this value with a result of necessity for a more sophisticated theory to obtain better agreement between the measured data and theory.

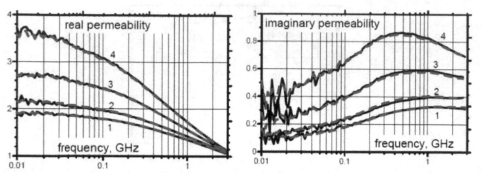

Figure 7. Black curves: the measured frequency dependencies of permeability of composites filled with milled iron powder (black lines), left – real permeability, right – imaginary permeability. Red curves – fitting of the measured data with theory (35). Volume fractions of inclusions are 15.0% (1), 17.7% (2), 23.6% (3), 30.3% (4) [39].

Recently, a new theory, which unites the MG and EMT approaches, has been proposed [46]. This theory allows for introducing the percolation threshold through a general quadratic equation, the same as the EMT, postulating two requirements to the solution. The On the one hand, the solution must be consistent with the LLL mixing rule (3) for the case of low intrinsic material parameter; on the other hand, it should satisfy the Odelevskiy equation (21) for the case of intrinsic material parameter tending to infinity. This produces a unique solution for the equation, which can be considered as a new mixing rule, which generalizes the EMT and MG mixing rules,

$$p\frac{\chi_{\text{eff}}}{\chi_{\text{incl}}} + \frac{1-p}{Dpn}\frac{1}{1/p - 1/p_c - 1/n/\chi_{\text{eff}}} = 1, \tag{35}$$

where D is the dimensionality of composite. Mixing rule (35) involves two fitting parameters: the effective form factor of inclusions, n, and the percolation threshold, p_c, that can be found from the concentration dependence of the effective permittivity. In fact, these parameters are related to peculiarities of morphology of composites, such as the distribution of inclusions in shape.

The derivation of Eq. (35) is based on the assumption that the spectral function has a single wide peak. It is shown [46] that Eq. (35) allows for a variation of these parameters over the ranges,

$$\sqrt{4/D} - 1 < n < 1/D,$$
$$\frac{1}{2}\left(Dn(1+n) - n\sqrt{D}\sqrt{D(1+n)^2 - 4}\right) < p_c < 1. \tag{36}$$

These conditions correspond to the case of nearly-spherical inclusions. Derivation of similar approach for composites filled with highly-elongated inclusions, such as thin platelets or fibers, must incorporate a spectral function comprising two separated peaks, which would require more sophisticated mathematical approaches. However, to develop such approach is not a challenging problem. As is shown above, the dilute limit approximation is sufficient for the analysis of microwave magnetic performance of such composites.

With the fitting parameters retrieved from the concentration dependence of permittivity, the intrinsic permeability of inclusions may be found from the measured effective permeability at each volume fraction of inclusions in the composite, as is described in [39]. An agreement of the data on the intrinsic permeability of inclusions found from different concentrations of inclusions provides an additional test for the validity of the mixing rule. It is found that the theoretical predictions agree closely with the measured microwave permittivity and permeability of composites filled with milled Fe powders [39], see Fig. 7. In the figure, the intrinsic permeability of inclusions was calculated for each concentration of inclusions, after which the average value was used to calculate the theoretical curve for each concentration. This is the reason for the noise observed in the theoretical curves in Fig. 7.

4. Composites with fibrous inclusions

4.1. Measured microwave permittivity of fiber-filled composites

Frequency dispersion of permittivity typically is not observed in composites over the microwave range. One of the rare examples of microwave dielectric dispersion is provided by composites filled with carbonized organic fibers. The conductivity of such fibers is much lower that that of metals. The thickness of the fibers is about a few microns, and their length can be on the order of several millimeters. The form factor of the fiber is very low, and the region of frequency dispersion may be at microwaves, as is seen in Fig. 8 [47].

Figure 8 shows the measured frequency dependence of permittivity for a composite filled with carbon fibers with length l=1.5 mm, thickness d=8 μm, and resistivity of 10 000 Ohm×cm. The volume fraction of the fibers in the composite is p=0.01%. The sample is a sheet polymer-based composite of thickness of less than 1 mm. Fibers are parallel to the sheet material plane, and they are distributed and oriented randomly in this plane. Experimental details are given in [47]. The frequency dependence of permittivity is of the Debye type. The low-frequency permittivity varies linearly with the volume fraction. The measured frequency and concentration dependences of permittivity agree well with the dilute limit approximation (4), written for the case under consideration as

$$\varepsilon_{\text{eff}} = \varepsilon_{\text{host}}\left(1 + p\kappa \frac{2\mathrm{i}\,\sigma/f - \varepsilon_{\text{host}}}{\varepsilon_{\text{host}} + n\left(2\mathrm{i}\,\sigma/f - \varepsilon_{\text{host}}\right)}\right), \tag{37}$$

where σ is the conductivity of the fibers, κ is a factor describing the averaged polarizability of inclusions, and n is the depolarization factor of the fibers,

$$n = \frac{d^2}{l^2}\ln\frac{l}{d}. \tag{38}$$

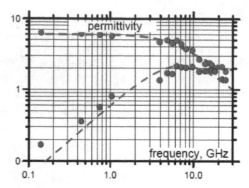

Figure 8. The measured frequency dependence of real (blue dots) and imaginary (red dots) permeability of a composite filled with carbon fibers of 1.5 mm in length with the resistivity of 10 000 Ohm×cm. The volume fraction of the fibers is 0.01% [47]. Curves are the result of fitting of measured data with the Debye dispersion law.

The value of κ =1/3 in Eq. (37) for the case under consideration, that is a product of the value of ½, which accounts for the isotropic in-plane orientations of the fibers in a sheet sample, and the value of 2/3, which accounts for cylindrical shape of fibers instead of ellipsoidal shape considered by the theories.

The type of the frequency dependence observed in fiber-filled composites is determined by the conductivity of fibers. In the general case, the dielectric dispersion curve is of the Lorentzian type with the parameters written as [48]

$$\chi_{st} = \frac{1}{3} \frac{\varepsilon_{host} p}{\ln l/d} \left(\frac{l}{d}\right)^2, \; f_{rel} \approx \frac{2\sigma}{\varepsilon_{host}} \left(\frac{d}{l}\right)^2 \ln l/d, \; f_{res} \approx \frac{c}{2l\sqrt{\varepsilon_{host}}} \quad (39)$$

The resonance of the permittivity arises from the half-wavelength resonance excited within the fibers.

Figure 9 shows the measured frequency dependence of permittivity for a composite filled with aluminum-coated fibers of 10 mm long [47]. The volume fraction the fibers is 0.01%. Due to high conductivity of the fibers, the frequency dependence of permittivity is of pronounced resonance (Lorentzian) type.

It is seen from Fig. 9 that the quality factor of the dielectric resonance is much lower than that predicted by Eq. (39). This is because Eq. (39) does not account for the radiation resistance of the fibers. The radiation resistance of a half-wavelength dipole is approximately 75 Ohm in the free space, which is much larger than the ohmic resistance of the fiber, and contributes dominantly to the quality factor of the resonance.

In fact, such composites behave as a kind of a metamaterial over the frequency range near the resonance, because they contain inhomogeneities, whose characteristic dimensions are close to the wavelength, and the principal features of their dielectric dispersion depend on the resonance scattering on the fibers. This is also evidenced by the facts that the measured permittivity is less than that produced by the MG mixing law, and that the radiation resistance makes a dominant contribution into the quality factor of the dielectric resonance. Rigorously, metamaterials cannot be described in terms of effective material parameters. However, an experimental observation of deviation of microwave performance of the composites from Fresnel law has required special measurement conditions, see [49] for details.

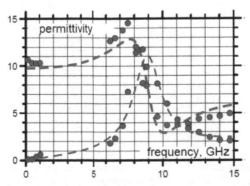

Figure 9. The measured frequency dependence of real (blue dots) and imaginary (red dots) permeability of a composite filled with aluminum-coated fibers 20 mm long. The volume fraction of the fibers is 0.01% [47]. Curves are the result of fitting of measured data with the Lorentzian dispersion law.

4.2. Theories for the effective properties of fiber-filled composites

A typical feature of fiber-filled composites is a low value of the percolation threshold: $p_c \propto d/l$, see [47] for the measured data. Although the percolation threshold is conventionally considered as a structure-dependent parameter, the dependence has been validated with composites based on a random mixture of conducting and non-conducting fibers, so that the dependence on agglomeration would be minimized. The standard EMT produces even lower values, $p_c \propto (d/l)^2$, and a large disagreement with the measured permittivity values at concentrations close to the percolation threshold can be observed. Mixing rules for the fiber-filled composites were primarily aimed at obtaining proper dependence of the percolation threshold on the aspect ratio of fibers.

Historically, the first theories describing the effective properties of fiber-filled composites have been suggested in [37] and [50]. However, the theory [37] describes the case of infinite conductivity of inclusions and is not suitable for describing of frequency dependences in metal-dielectric composites. Theory [50] is a modification of the EMT. It is based on the assumption of strong anisotropy of the effective medium in the vicinity of a particular fiber, which results in the equations

$$\frac{p}{3}\left(\frac{\varepsilon_{\text{eff},||}}{\varepsilon_{\text{eff},||} + n\left(\varepsilon_{\text{incl}} - \varepsilon_{\text{eff},||}\right)} + \frac{2\varepsilon_{\text{eff},\perp}}{\varepsilon_{\perp} + n_{\perp}\varepsilon_{\text{incl}}\left(\varepsilon_m - \varepsilon_{\text{eff},\perp}\right)} \right) + \frac{\left(1 - p\right)\varepsilon_{\text{eff},\perp}}{\varepsilon_{\perp} + n\left(\varepsilon_{\text{host}} - \varepsilon_{\text{eff},\perp}\right)} = 1, \qquad (40)$$

$$\frac{p}{3}\left(\frac{\left(\varepsilon_{\text{eff,II}}-\varepsilon_{\text{incl}}\right)}{\varepsilon_{\text{eff,II}}+n\left(\varepsilon_{\text{incl}}-\varepsilon_{\text{eff,II}}\right)}+\frac{2\left(\varepsilon_{\text{eff,}\perp}-\varepsilon_{\text{incl}}\right)}{\varepsilon_{\perp}+n_{\perp}\left(\varepsilon_{\text{incl}}-\varepsilon_{\text{eff,}\perp}\right)}\right)+\frac{\left(1-p\right)\left(\varepsilon_{\text{eff,}\perp}-\varepsilon_{\text{host}}\right)}{\varepsilon_{\perp}+n\left(\varepsilon_{\text{host}}-\varepsilon_{\text{eff,}\perp}\right)}=0. \tag{41}$$

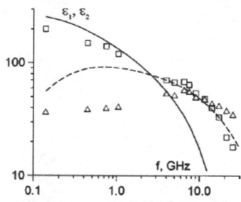

Figure 10. Dots: the measured dielectric dispersion curves for a composite filled with carbon fibers 1.8 mm long with the resistivity of 1400 Ohm×cm. The volume concentration of the fibers is 0.52%. Curves: calculation by Eq. (42) [47].

The two equations (40) and (41) are used for the search of two values of the effective permittivity, $\varepsilon_{\text{eff,II}}$ and $\varepsilon_{\text{eff,}\perp}$, in the directions parallel and perpendicular to a fiber, respectively; n is given by Eq. (38), $n_{\perp}=(1-n)/2$, and the observed effective permittivity of the composite is found by averaging of $\varepsilon_{\text{eff,II}}$ and $\varepsilon_{\text{eff,}\perp}$. The value of the randomization factor κ, defined in the Subsection 2.1, equals to 1/3, and is substituted in Eqs. (40) and (41). The theory [50] predicts correctly the dependence of the percolation threshold on the aspect ratio of fibers, but disagrees with the dilute limit approximation, and, therefore, with the measured permittivity of composites at low concentrations of fibers.

Theory [51] allows for a better quantitative agreement over a wide range of volume fractions below the percolation threshold. The theory is based on the assumption that, in the vicinity of a particular fiber, the permittivity of effective medium ε^* is a function of the distance from to fiber z:

$$\begin{aligned}\varepsilon^*\left(z\right)&=\varepsilon_{\text{host}}+\left(\varepsilon_{\text{eff}}-\varepsilon_{\text{host}}\right)\left(z/ax\right) \quad \text{at} \quad z<ax \\ \varepsilon^*\left(z\right)&=\varepsilon_{\text{host}} \quad \text{at} \quad z\geq ax\end{aligned} \tag{42}$$

where x is a parameter of the theory. This assumption results in the EMT equation written as

$$\frac{\varepsilon_{\text{incl}}}{3\varepsilon_{\text{eff}}}\frac{p}{1+\left(d^2\varepsilon_{\text{incl}}/l^2\varepsilon_{\text{host}}\right)\ln\left(1+\left(lx\varepsilon_{\text{host}}\right)/\left(d\varepsilon_{\text{eff}}\right)\right)}+3\frac{\varepsilon_{\text{host}}-\varepsilon_{\text{eff}}}{\varepsilon_{\text{host}}+2\varepsilon_{\text{eff}}}=0, \tag{43}$$

which is conventionally used in microwave studies of fiber-filled composites, see, e.g., [52].

There is lack of measured data on the frequency-dependent dielectric performance of fiber-filled composites near the percolation threshold in the literature; one of examples of the data is given in Fig. 10 [45]. It is seen from the figure that the EMT predicts a gradual shift of the loss peak. The measured dielectric loss peak differs from that predicted by the theory. In contrast, the measured variation of the loss peak appearing as approaching the percolation threshold looks like a rise of the low-frequency loss level, with a well-defined trace of the loss peak associated with individual fibers.

This difference between the theory and measurements may be associated not with the geometrical distribution of shapes of conducting clusters, as the percolation theory suggests, but with other low-frequency loss mechanisms near the percolation threshold. For example, imperfect electric contacts between fibers comprising a conductive cluster may contribute to the low-frequency loss [20]. The conductivity of such contacts must be much lower than the conductivity of the fibers. Therefore, imperfect contacts may result in a large low-frequency shift of dielectric loss. Because of a low value and wide distribution of the conductivity of contacts, this loss forms a very smooth dispersion curve, which is seen in Fig. 10.

The same may be true for composites filled with carbon black or carbon nanotubes, which are known to have the percolation type of frequency dispersion at microwaves, see, e.g., [20]. Dielectric loss appearing at low frequencies might be associated with very prolate conductive clusters, if the conductivity of clusters is on the same order of magnitude as the conductivity of inclusions. Account for the imperfect contacts would allow for more realistic assumptions on the shape of conducting clusters.

In principle, the effect of contacts may be understood as a presence of a comparatively low-conductive shell covering the surface of conducting inclusions. For an individual inclusion, the presence of such a shell leads to a low-frequency shift of the loss peak, without change in its shape. To get an agreement with the measured data, a distribution of these conductivities should be included in the model. By the analogy to the Cole-Cole dispersion law, such distribution would results in the power frequency dependences of the permittivity. Hence the difference between measured critical indices and universal values derived from geometrical considerations can be observed, but there is no theory explaining and quantifying such phenomena in the literature.

The available data of the microwave permeability of composites filled with magnetic fibers are consistent with the dilute limit approximation [53, 54].

5. Conclusions

The problem of describing of the effective permittivity as a function of concentration of inclusions in a metal-dielectric mixture is well studied. However, newly developed mixing rules still appear in the literature. This means that the solution for the problem is not satisfactory to some extent, and is typically related to the description of frequency dependences of material parameters.

The above consideration allows for determining the validity limits of various mixing rules. These limits are dependent on the difference between the susceptibilities of inclusions and the host matrix, and on the elongation of inclusions.

For microwave permeability, the difference is typically not high, and the effective properties of composites are well described by the MG mixing rule. For lower frequencies, the intrinsic permeability may be high, and a more sophisticated mixing rule may be needed. For composites containing platelet and fibrous magnetic inclusions, the microwave permeability is described by the dilute limit approximation. The same is true for composites with dielectric fibers.

For microwave permittivity of a metal-dielectric mixture, the difference is typically large, and the effective properties are determined by the morphology of the composite. But fitting of measured data to the theoretical results is typically rather simple, because the frequency dispersion of permittivity is a rare occasion at microwaves. In metal-dielectric composites, the region of frequency dispersion is located at much higher frequencies, as can be estimated from typical conductivity of metals and feasible dimensions of inclusions.

For simultaneous modeling of the permittivity and permeability of composites with conducting inclusions, sophisticated mixing rules are unavoidable, with an account for a distribution of inclusions in shape. This case is the most difficult, because both concentration and frequency dependences of material parameters may be non-trivial.

Author details

Konstantin N. Rozanov
Institute for Theoretical and Applied Electromagnetics, Russian Academy of Sci., Moscow, Russia,

Marina Y. Koledintseva
Missouri University of Science and Technology, Rolla, MO, USA

Eugene P. Yelsukov
Physical and Technical Institute, Ural Branch of Russian Academy of Sci., Izhevsk, Russia

Acknowledgement

K. Rozanov acknowledges the partial financial support of the work from the RFBR, grants no. 12-02-91667 and 12-08-00954. M. Koledintseva acknowledges the partial support by the U.S. NSF Grant No. 0855878. The authors also thank Alexei Koledintsev for his assistance and valuable comments regarding technical English writing.

6. References

[1]   Koledintseva MY, Rozanov KN, Drewniak JL (2011) Engineering, modeling and testing of composite absorbing materials for EMC applications, In: Adv. in Composite

Materials – Ecodesign and Analysis, ed. B. Attaf, InTech, ISBN 978-953-307-150-3, Ch. 13, pp. 291–316.

[2] Garnett JCM (1904) Colours in Metal Glasses and in Metallic Films. Phil. Trans. R. Soc. Lond. 203: 385–420.

[3] Bruggeman DAG (1935) Berechnung Verschiedener Physikalischer Konstanten von Heterogenen Substanzen. Ann. Phys. (Leipzig) 24: 636–679.

[4] Landau LD, Lifshitz EM (1984) Electrodynamics of Continuous Media, Pergamon, 474 p.

[5] Looyenga H (1965) Dielectric Constants of Heterogeneous Mixtures. Physica. 31: 401–406.

[6] Born M, Wolf E (1986) Principles of Optics. 6 Ed., Pergamon, 854 p.

[7] Polder D, van Santen JH (1946) The Effective Permeability of Mixtures of Solids. Physica. 12: 257–271.

[8] Bergman DJ, Stroud D (1992) Physical Properties of Macroscopically Inhomogeneous Media. Solid. State Phys. 46: 147–269.

[9] Hashin Z, Shtrikman S (1962) A Variational Approach to the Theory of the Effective Magnetic Permeability of Multiphase Materials J. Appl. Phys., 33: 3125–3131.

[10] Sareni B, Krahenbuhl L, Beroual A, Brosseau C (1996) Effective Dielectric Constant of Periodic Composite Materials. J. Appl. Phys. 80: 1688–1696.

[11] Sihvola AH (1999) Electromagnetic Mixing Rules and Applications. IET, 284 p.

[12] Torquato S, Hyun S (2001) Effective Medium Approximation for Composite Media: Realizable Single-Scale Dispersions. J. Appl. Phys. 89: 1725–1729.

[13] Reynolds JA, Hough JM (1957) Formulae for Dielectric Constant of Mixtures. Proc. Phys. Soc. B 70: 769–775.

[14] Lagarkov AN, Rozanov KN (2009) High-Frequency Behavior of Magnetic Composites. J. Magn. Magn. Mater. 321: 2082–2092.

[15] Day AR, Grant AR, Sievers AJ, Thorpe MF (2000) Spectral Function of Composites from Reflectivity Measurements. Phys. Rev. Lett. 84: 1978–1981.

[16] Theiss W (1996) The Dielectric Function of Porous Silicon – How to Obtain It and How to Use It. Thin Solid Films. 276: 7–12.

[17] Ghosh K, Fuchs R (1988) Spectral Theory for Two-Component Porous Media. Phys. Rev. B: 38 5222–5236.

[18] Osipov AV, Rozanov KN, Simonov NA, Starostenko SN (2002) Reconstruction of Intrinsic Parameters of a Composite from the Measured Frequency Dependence of Permeability. J. Phys.: Condens. Matter 14: 9507–9523.

[19] Goncharenko AV, Lozovski VZ, Venger EF (2000) Lichtenecker's Equation: Applicability and Limitations. Optics. Commun. 174: 19–32.

[20] Liu L, Matitsine S, Gan YB, Chen LF, Kong LB, Rozanov KN (2007) Frequency Dependence of Effective Permittivity of Carbon Nanotube Composites. J. Appl. Phys. 101: 094106.

[21] Jonscher AK (1983) Dielectric Relaxation in Solids, Chelsea Dielectrics Press, 1983, 380 p.

[22] Mattei JL, Le Floc'h M (2003) Percolative Behaviour and Demagnetizing Effects in Disordered Heterostructures. J. Magn. Magn. Mater. 257: 335–345.

[23] Pitman KC, Lindley MW, Simkin D, Cooper JF (1991) Radar Absorbers: Better by Design. IEE Proc. F – Radar and Signal Processing 138: 223–228.

[24] Miyasaka K, Watanabe K, Jojima E, Aida H, Sumita M, Ishikawa K (1982) Electrical Conductivity of Carbon-Polymer Composites as a Function of Carbon Content. J. Mater. Sci. 17: 1610–1616.

[25] Duan HL, Karihaloo BL, Wang J, Yi X (2006) Effective Conductivities of Heterogeneous Media Containing Multiple Inclusions with Various Spatial Distributions. Phys. Rev. B 73: 174203.

[26] Gao L, Gu JZ (2002) Effective Dielectric Constant of a Two-Component Material with Shape Distribution. J Phys. D – Appl. Phys. 35: 267–271.

[27] Goncharenko AV (2003) Generalizations of the Bruggeman Equation and a Concept of Shape-Distributed Particle Composites. Phys. Rev. E 68: 041108.

[28] Koledintseva MY, Chandra SKR, DuBroff RE, Schwartz RW (2006) Modeling of Dielectric Mixtures Containing Conducting Inclusions with Statistically Distributed Aspect Ratio. PIER 66: 213–228.

[29] Spanoudaki A, Pelster R (2001) Effective Dielectric Properties of Composite Materials: The Dependence on the Particle Size Distribution. Phys. Rev. B 64: 064205.

[30] Koledintseva MY, DuBroff RE, Schwartz RW, Drewniak JL (2007) Double Statistical Distribution of Conductivity and Aspect Ratio of Inclusions in Dielectric Mixtures at Microwave Frequencies. PIER 77: 193–214.

[31] Koledintseva MY, DuBroff RE, Schwartz RW (2009) Maxwell Garnett Rule for Dielectric Mixtures with Statistically Distributed Orientations of Inclusions. PIER 99 131–148.

[32] Lichtenecker K (1926) Die Dielektrizitätskonstante Natürlicher und Künstlicher Mischkörper. Physikal. Z. 27: 115–158.

[33] McLachlan DS, Priou A, Chenerie I, Isaak E, Henry F (1992) Modeling the Permittivity of Composite Materials with a General Effective Medium Equation. J. Electromagn. Waves Appl. 6: 1099–1131.

[34] Doyle WT, Jacobs IS (1990) Effective Cluster Model of Dielectric Enhancement in Metal-Insulator Composites. Phys. Rev. B 42: 9319–9327.

[35] Sheng P (1980) Theory for the Dielectric Function of Granular Composite Media. Phys. Rev. Lett. 45: 60–63.

[36] Musal HM, Hahn HT, Bush GG (1988) Validation of Mixture Equations for Dielectric-Magnetic Composites. J. Appl. Phys. 63: 3768–3770.

[37] Doyle WT, Jacobs IS (1992) The Influence of Particle Shape on Dielectric Enhancement in Metal-Insulator Composites. J. Appl. Phys. 71: 3926–3936.

[38] Odelevskiy VI (1947) The Calculation of Generalized Conductivity of Heterogeneous Systems. Ph. D. Thes., Moscow, 110 p.

[39] Rozanov KN, Osipov AV, Petrov DA, Starostenko SN, Yelsukov EP (2009) The Effect of Shape Distribution of Inclusions on the Frequency Dependence of Permeability in Composites. J. Magn. Magn. Mater. 321: 738–741.

[40] Li M, Feteira A, Sinclair DC, West AR (2007) Incipient Ferroelectricity and Microwave Dielectric Resonance Properties of $CaCu_{2.85}Mn_{0.15}Ti_4O_{12}$ Ceramics. Appl. Phys. Lett. 91: 132911.

[41] Rozanov KN, Li ZW, Chen LF, Koledintseva MY (2005) Microwave Permeability of Co_2Z Composites. J. Appl. Phys. 97: 013905.

[42] Lagarkov AN, Osipov AV, Rozanov KN, Starostenko SN (2005) Microwave Composites Filled with Thin Ferromagnetic Films. Part I. Theory. Proc. Symp. R: Electromagn.. Mater., 3rd Int. Conf. on Mater. Adv. Technol. (ICMAT 2005), Jul. 3–8, 2005, Singapore, pp. 74–77.

[43] Acher O, Dubourg S (2008) Generalization of Snoek's Law to Ferromagnetic Films and Composites. Phys. Rev. B 77: 104440.

[44] Snoek JL (1948) Dispersion and Absorption in Magnetic Ferrites at Frequencies above 1 Mc/s. Physica 14: 207–217.

[45] Acher O, Adenot AL (2000) Bounds on the Dynamic Properties of Magnetic Materials. Phys. Rev. B 62: 11324.

[46] Rozanov KN, Koledintseva MY, Drewniak JL (2012) A Mixing Rule for Predicting of Frequency Dependence of Material Parameters in Magnetic Composites J. Magn. Magn. Mater. 324: 1063–1066.

[47] Lagarkov AN, Matytsin SM, Rozanov KN, Sarychev AK (1998) Dielectric Properties of Fiber-Filled Composites. J. Appl. Phys. 84: 3806–3814.

[48] Matitsine SM, Hock KM, Liu L, Gan YB, Lagarkov AN, Rozanov KN (2003) Shift of Resonance Frequency of Long Conducting Fibers Embedded in a Composite. J. Appl. Phys. 94: 1146–1154.

[49] Vinogradov AP, Machnovskii DP, Rozanov KN (1999) Effective Boundary Layer in Composite Materials. J. Communic. Technology Electr. 44: 317–322.

[50] Lagarkov AN, Sarychev AK, Smychkovich YR, Vinogradov AP (1992) Effective Medium Theory for Microwave Dielectric Constant. J. Electromagn. Waves Appl. 6: 1159–1176.

[51] Lagarkov AN, Sarychev AK (1996) Electromagnetic Properties of Composites Containing Elongated Conducting Inclusions. Phys. Rev. B 53: 6318–6336.

[52] Makhnovskiy DP, Panina LV, Mapps DJ, Sarychev AK (2001) Effect of Transition Layers on the Electromagnetic Properties of Composites Containing Conducting Fibres. Phys. Rev. B 64: 134205.

[53] Liu L, Kong LB, Lin GQ, Matitsine S, Deng CR (2008) Microwave Permeability of Ferromagnetic Microwires Composites/Metamaterials and Potential Applications. IEEE Trans. Magn. 44: 3119–3122.

[54] Han MG, Liang DF, Deng LJ (2011) Fabrication and Electromagnetic Wave Absorption Properties of Amorphous Fe79Si16B5 Microwires. Appl. Phys. Lett. 99: 082503.

Chapter 4

The Chosen Aspects of Materials and Construction Influence on the Tire Safety

Pavel Koštial, Jan Krmela, Karel Frydrýšek and Ivan Ružiak

Additional information is available at the end of the chapter

1. Introduction

Security of the road transport depends on the quality of basic and applied research concerning materials and internal construction of tires. The design shapes and material properties characterized by low hysteretic losses as well as a construction have an influence on the driving comfort, adhesion, wear resistance and fatigue resistance. Thick fibre reinforced composites are used extensively in rubber products such as tires and conveyer belts. Generally, the reinforced parts of rubber products on a sub macroscopic level are highly heterogeneous and anisotropic because they are composed of rubber compounds, and textile and steel cords. Rubber compounds consist of natural or synthetic rubber, carbon black, curing agents, cure accelerators, plasticizers, protective agents and other ingredients.

First, the properties of the different parts of the studied tire will be outlined. A bead is a part of the tire, which fixes to the rim. The bead consists of a steel bead wire, a core, a bead filling and a carcass. They help to transmit loading and breaking.

Particularly we will focus our attention on the influence of breaker angle on tire deformation and potential risks resulting from improper breaker construction.

Experimental results of tread and side wall deformation (influenced by rubber blend as well as a breaker construction) measured independently by both line laser and Aramis system are compared with those obtained by computer simulation in Abaqus environment. The tire tread contributes to a good road grip and water expulsion, the multi-ply steel belt optimizes the directional stability and rolling resistance, the steel casing substantially determines the driving comfort, the inner-liner makes the tire airtight, the sidewall protects from lateral scuffing and the effects of the weather, the bead core ensures the tire sits firmly on the rim, and bead reinforcement promotes directional stability and a precise steering response.

The high security and long life of a tire can be assured only by its correct assignment to the particular type of vehicle and automobile as show Figure 1 (tires only for road operation, for off-road, combined operation as well as for summer or winter conditions). Tires are divided by type of tire-casing on radial, diagonal, bias-belted and special tire.

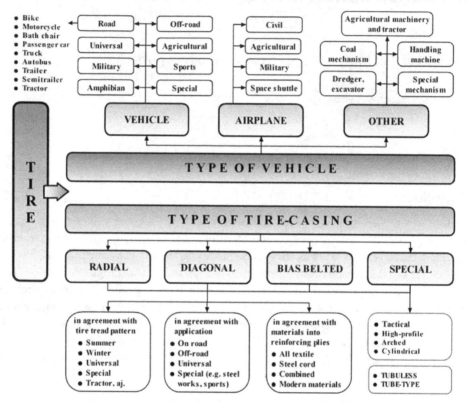

Figure 1. Type of tire by vehicle and construction

Radial tires can be considered as pressure vessels with a maximum pressure given by the particular type of tire. A tire can be generally considered as a statically and dynamically loaded automobile element. The structured view of a tire is apparent from the Figure 2.

The function of wheels with tires is not only to align a car reliably. As can be seen in more detail from the Figure 3 there are more requirements on tires. The main operating requirements on car tires are that car wheels should be as light as possible and at the same time tough, statically and dynamically balanced.

The main requirements on tires are, apart from other things, high wear resistance, optimal deformation characteristics, low rolling resistance, high operational life and safeness, etc. Wheels with tires must meet particular functional requirements given by parameters of tires which affect the running properties of the car, i.e. affect their dynamic behavior (car maneuverability, stability, acceleration, deceleration, driving comfort, etc.).

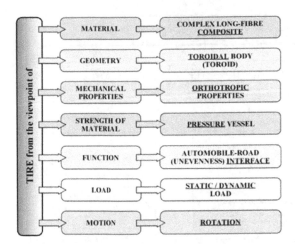

Figure 2. Definition of tire from various viewpoints

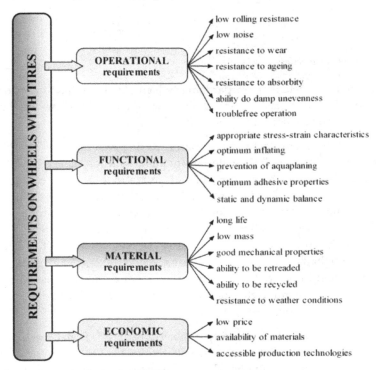

Figure 3. Basic requirements of wheels with tires

Tire safety is passive and active - see Figure 4. Passive safety depends on the quality of the production of a tire casing, the applied technology and used materials and in the case of computational modeling also on the accuracy of the performed calculations and appropriate choice of the computing algorithm.

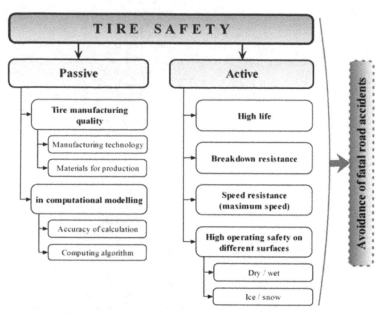

Figure 4. Viewpoints of tire safety

Requirements on active safeness are particularly high running safety on various types of road surfaces, breakdown resistance, speed resistance and high life of materials used for the production of tires, namely reinforcing materials.

Tire life is affected by many factors (e.g. manufacturing way of a tire, its operation and handling, storage conditions of base materials for tire production etc.) while it's assumed an ideal adhesive bond among the rubber elements in matrix (e.g. an interface between tire tread and textile overlap belt) and the cord reinforcing and rubber drift inside tire carcass, and belts is assumed in this cases. For long life tires must resist during operating to surrounding effects, to negative effects of operation and to other effects, which could lead to wear and degradation processes as are e.g. delaminations. The aim is to avoid fatal road accidents which might be caused by tire casing defects either by neglecting operating conditions of tires (depth of tire tread pattern, tire inflation pressure, use of inappropriate tires with a different structure, etc.) or by bad vulcanization during the manufacturing process creating delaminations.

The tire is during the operation exposed to combined loading as from a mechanical (statical, dynamic) as a temperature point of view (local heating in subzones, overall heating in the tire-tread area permeating into the tire during breaking). Also this has to be considered in defining tire safety at high speeds. For this reason tires and wheels as a unit are modified from the structural point of view, particularly for special army vehicles where even a sudden drop of pressure does not put an end to the operating capability of the vehicle (system with a central collar providing circular indexing of the casing with respect to the wheel rim).

New features are introduced for high speeds, e.g. electronic systems which warn the drivers in the case a gradual drop of the tire pressure or adjusting systems for inflation based on the temperature load of the tire casing. Each manufacturer protects the results of his developments and patents considering them as private „know-how". Consequently all new information is only very scarcely available.

In the material point of view a rubber blend could be considered as a composite material which consists of a matrix, a filler and a mesophase.

The dynamic – mechanical properties of such composites can also be described by an elastic, viscous modulus and a loss factor, (Simek 1987, Sepe 1998, Wang 1998, Schaefer 1994, Murayama 1978, Ferry 1980, Jančíková 2006, Jančíková & Švec 2007 and Jakubíková et al 2007). Such properties depend, in turn, on the operating temperature and frequency of the external excitation. Biodegradability of polymers has been studied in (Jakubíková et al 2007).

In a molecular scale, the mechanical properties of rubber blends are influenced mainly by the structure of the blend. The interaction between a matrix and filler plays the most important role and this role is closely connected with dependency of E and G modulus to an applied load or a frequency (both functions have falling tendency and this phenomenon is called Payne effect) (Payne 1965). The polymer in the network loses its identity, and behaves like a filler. The loss factor E'' depends on the dissolution and regeneration speed of the network. It is reflected on the decreasing trend of complex Young's modulus dependence versus increasing sample loading. The values of elastic modulus (E') for vulcanizates without fillers are not changed with the increasing of dynamic deformation (Payne 1965, Medalia 1978 and Maier & Gand Göritz 1996).

The blend properties are characterized by the following parameters: T_g is a glass transition temperature, it is influenced by the silica filler, and for not filled rubber it is approximately - 40°C. The phase shift between stress and strain is the loss angle. It is postulated that the loss factor, represented by *tan δ*, in the temperature span -10°C to 5°C (frequency 10000 Hz) characterizes the adhesion of the tire on a wet road. In the span 60°C to 80°C (at the frequency 100 Hz) the course of *tan δ*, characterizes the rolling resistance (http://ao4.ee.tut.fi/pdlri).

The breaker angle (see below) also influences the security of a tire as well as the driving comfort and stiffness of a tire. The driving properties of the tire as a whole could be substantially improved by optimizing such angle of steel wires of the breaker.

2. The tire description

The work of the authors over a long period of time is devoted to radial tires. The automobile radial tire consists (e.g. cross-sections of selected tire 165 R13 Matador are on the Figures 5 and 6, in detail on the Figures 6 below and 7) of rubber parts and composite structure parts (Figure 8) with textile cords (especially PA 6.6 and PES textile fibers are used) and steel-cords into tire tread as reinforcements.

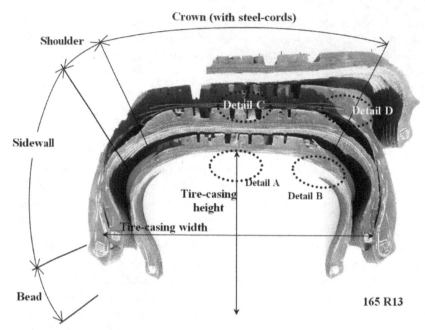

Figure 5. Cross sections of tire-casing

The composite structure parts applied into radial tires (Figure 8) are:

- Textile tire carcass;
- Textile overlap belt;
- Steel-cord belt.

These structures of tire have got:

- Different cord-angle (e.g. for steel belt applied angle 21-27° into radial tire for passenger car);
- Material of cords (steel, textile, Kevlar, combine);
- Shape and construction of cord (wire, wire strand);
- Numbers of layer (single-layer or multi-layer).

So a tire has got characteristic specific deformation properties.

One construction of tire is used for passenger cars, other constructions for trucks, off-highway cars and sports cars. The tires for air transportation, agricultural vehicle, mining machine and other vehicles have got complicated structured in comparison radial tires for passenger cars. The tire structures are differentiated by numbers of reinforcing plies into belt tire, construction of belts, materials and cord-angles, geometry parameters of tire, width of belts etc. These aspects are influenced on final behavior of tires, namely deformation characteristics of tires. It is possible increase of resistance of tire to some degradation processes by suitable tire construction.

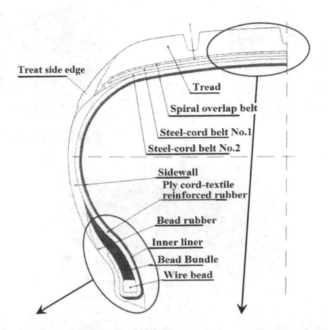

Treat side edge

Tread

Spiral overlap belt

Steel-cord belt No.1
Steel-cord belt No.2

Sidewall
Ply cord-textile
reinforced rubber

Bead rubber

Inner liner

Bead Bundle

Wire bead

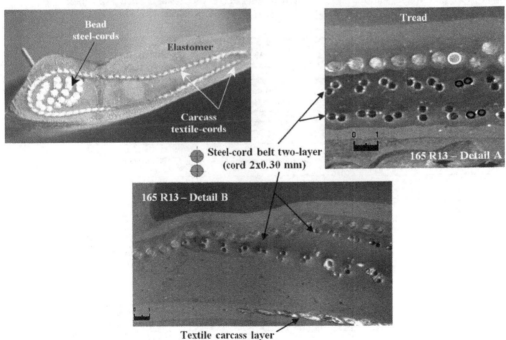

Bead
steel-cords

Elastomer

Carcass
textile-cords

Steel-cord belt two-layer
(cord 2x0.30 mm)

Tread

165 R13 – Detail A

165 R13 – Detail B

Textile carcass layer

Figure 6. Structure of the tire 165 R 13 [based on Matador] with microstructure of reinforcing plies detail A in the area of tire crown and detail B at the end of steel-cord belt (below)

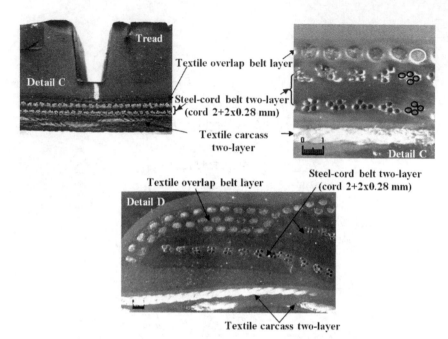

Figure 7. Detail C in the middle of tire crown and detail D at the end of the belt layers (below) of different radial tire

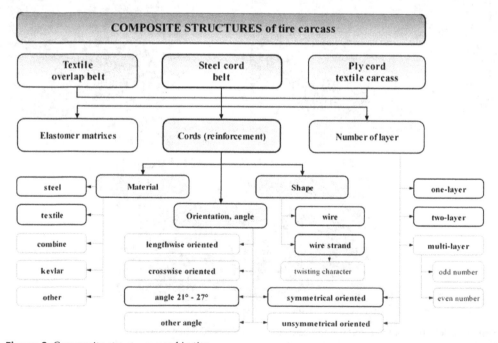

Figure 8. Composite structures used in tire

Two-layer steel-cord belt is used in radial tire *165 R13 Matador* with construction of cord *2x0.30 mm* with texture *961* (number of cord over meter width of belt). The cord angle is *23°*, the layers are symmetrical.

Structure – cord orientations of radial tire *22.5"* for truck vehicles presented in the Figure 9 as an example.

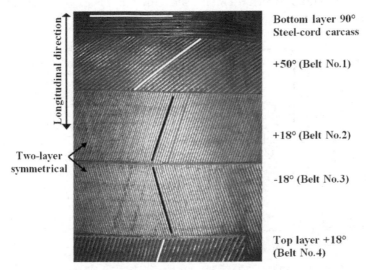

Figure 9. Structure of truck tire in the middle of tire crown

Steel-cords can be in form of thin wire or wire strand with different constructions. High-strength steels are used exclusively for steel-cord production and good adhesive bond between rubber and cords required. Steel-cord surfaces are modified by chemical-thermal treatment (braze or copperier, Figure 10) to achieve the best adhesive bond of a steel cord and rubber and get it corrosion resistant. The substantial factor, which expressive influence on coherence of whole tire, is good adhesive bonds between reinforcement materials and rubber parts of tire.

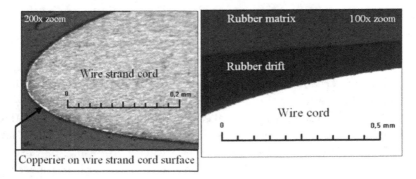

Figure 10. Steel-cord surface and interface between cord-rubber drift-rubber matrix

The tire steel-cords are exposed to various chemical and thermal influences (Figure 11) during cyclic loading states by tensile-compression in tire loading processes. Account on this the adhesive bond is more exposed to be damaged than the basic materials (steel, textile and rubber). The aggressive environment (e.g. action of salts in winter) activates the corroding process on steel-cord surfaces that can lead to decreasing of the adhesion between reinforcement-and-matrix, which demonstrates itself by negative changes in material properties of steel-cord belts and such of whole tire too.

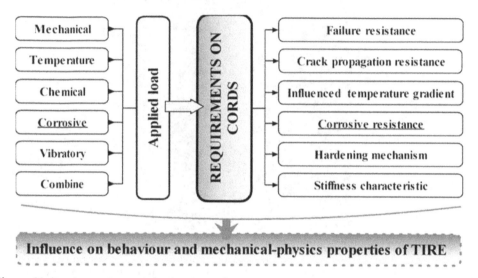

Figure 11. Requirements on reinforcing tire cords

In addition if the tire is in use is defected in tire crown (e.g. defect caused by sharp object as a nail and after the repair is placed back into operation) the initiation of corrosion with faster process is being assumed. Consequently this can lead to gradual or sudden failure of the steel-cords and bonds of steel-cord and rubber with a serious car accident as a final consequence.

Any damage in the area of tire crown, namely into steel-cord belt, is perilous.

3. Degradation processes of tires

Tires are subject to internal and external effects which can more or less cause limit states leading to degradation processes Figure 12. Ones of them marked as very dangerous and unacceptable tire casing damage are so-called separations and delaminations (Figure 13 left). Breakdown or damage is not necessary only at the border of single layer e.g. between the layers of steel-cord belt plies in the tread of tire casing, but also between rubber matrix and reinforcing cords. Delamination between rubber drift-rubber matrix on Figure 13 right.

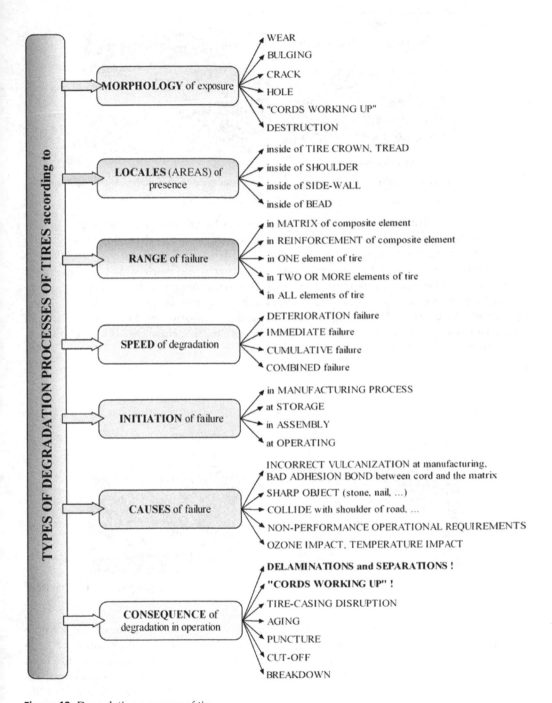

Figure 12. Degradation processes of tire

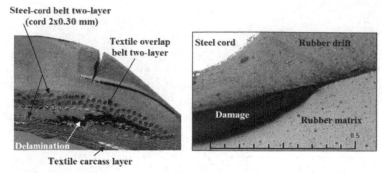

Figure 13. Factitious delamination between belt plies and damage between rubber drift-rubber matrix (right)

Root cause of mentioned degradation processes can be caused by using of low quality materials for the tire in manufacturing process, their incorrect storage leading to early aging especially at rubber compositions, not keeping optimal manufacturing conditions – vulcanization, as well as by the influence of incorrectly pressurized tire and damaged adhesive bond between cords-matrix and belt plies etc. Damaged adhesive bond is greatly decreasing of tire safety during the operation of a vehicle at high speeds. This has a significant influence to the quality of the tire casing expressing by lowering the level of usage (decreasing speed index) or leading to the catastrophic situations. In every case it is mandatory to avoid these premature limiting states.

In tires can be caused:

- Cords release leading to the cords working-up;
- Separations – delaminations;
- Combined wear;
- Total breakdown.

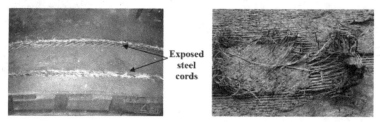

Figure 14. Extreme wear of tire on tread surface; and "cords working-up" (right-photo by Prof. Janíček, VÚT Brno, Czech Republic)

Results of wear due to adverse change (Figure 14) of tire casing surfaces are gave in impairment of mechanical-physics properties of whole tire. This will be influenced the incoming behavior of tire in operation and related interfaces between of tire and surroundings. Particularly dangerous is creation of failure in such places where initiation is not assumed to be caused by impairment of the surface. Structural changes in a part of tire as composites are not only responsible for the impairment of its properties but also of its geometry which can initiate vibrations leading to loss of the part's functional ability of whole vehicles (automobiles).

Wear can be of various character (development, place, form, appearance) and leads to the failure of the tire (Figure 15a and 15b). The task of prediction is to find ways how to reduce wear and to postpone initiation of dangerous degradation processes such as delamination and separation and to focus on the removal of initiators of these degradation processes.

Wear due to adverse changes of surfaces results in impairment of properties and behavior of parts. Particularly dangerous is failure in such places where initiation is not assumed to be caused by impairment of the surface. Structural changes in a part are not only responsible for the impairment of its mechanical properties but also of its geometry which can initiate vibrations leading to loss of the part's functional ability. All this resulted in environmental and economical losses.

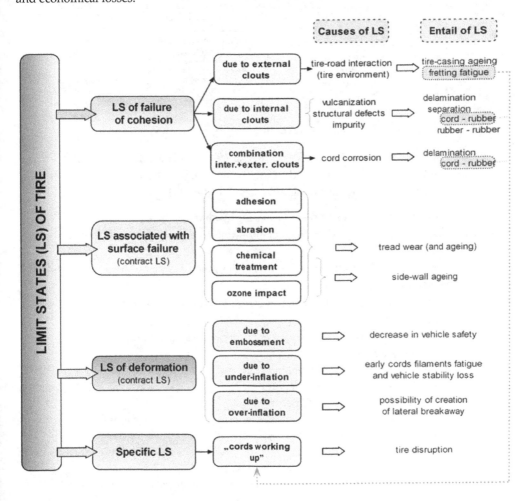

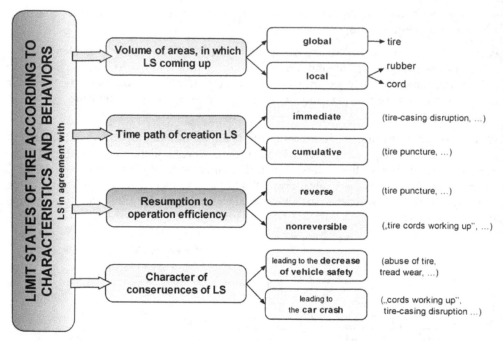

Figure 15. a. Limit states of tire; b. Limit states of tire

Tires must resist during operating to surrounding effects, to negative effects of operation and to other effects, which could lead to wear and degradation processes as are e.g. delamination. Resistance to the following effects is considered (Figure 16):

- Puncture – capability of tires to resist puncture by sharp objects;
- Cut-through – capability of tires (especially of the tread and sidewall) to resist contact with sharp objects;
- Breakdown – capability of tires to resist damage during short-term loading by concentrated forces;
- Fatigue – capability of tires to resist material fatigue and defects in consequence of repeated loading cycles;
- Separation and delamination – capability of structural tire components to maintain integrity of the system during operation;
- Humidity – tire elements must be able to resist degradation by contact with water;
- Ozone influence – capability of tires and of theirs components to resist degradation caused by ozone present in atmosphere;
- Temperature – tire components must be able to resist high and low ambient temperatures and also consequences of contact with the road;
- Chemicals – capability of tires and theirs components to resist degradation caused by chemicals (in winter – influence of salt solutions);
- Corrosion processes – capability of tire reinforcing cords to resist corrosion, etc.

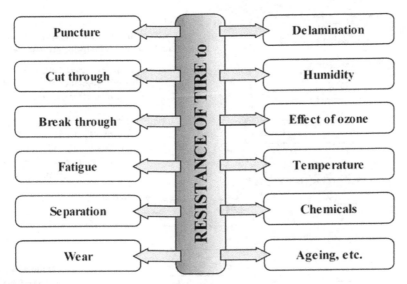

Figure 16. Basic requirements on the tire resistance

An appropriate design will help to increase resistance of the tire to certain degradation processes such as e.g. corrosive attacks initiated by local damage of the tire in cases where damaged are the steel cord reinforcements. Design optimization aimed at resistance to degradation and at achievement of longer life can be well performed by computer modelling. The computer modelling has reached such a level that it can work with a great amount of input data which represent the initiation of degradation effects on such a complicated technical object as a tire.

It is important to design such structure that the tire would be as much resistant to any degradation type as possible. These are required complex approach to experiments and computation of tire from macrostructure and microstructure too. It is necessary to have a good knowledge about:

- Structure of tire-casing;
- Material parameters of matrixes and reinforcements (steel-belt);
- Adhesive bonds cord-rubber, which obtained by metallography observation of reinforcements-matrix transit;
- Influence of degradation processes – corrosion effect on composite materials from micro and macrostructure point of view.

4. Testing of tires

It is necessary to run tests of tires as a whole, as shown in the Figure 17, and tests of individual tire casing components, purposely separated parts etc. This is how an overview which structural modifications can lead to an increase of the level of safety criteria, increase of resistance, life etc. can be obtained.

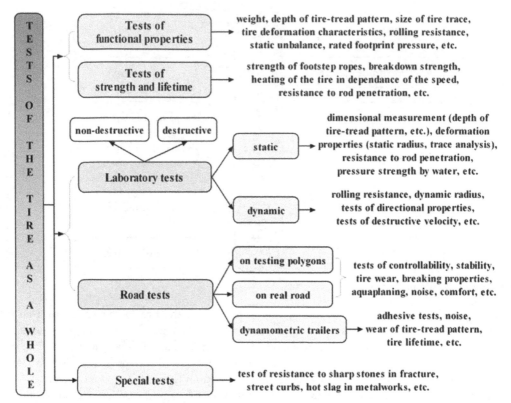

Figure 17. Tests of the tire as a whole

Figure 18. Static adhesor with detail of contact patch

Also basic statical deformation characteristics of tires can be obtained from a device called statical adhesor (Figure 18), which is available to author. The statical adhesor also enables measurement of data from the contact surface under defined conditions – shape of obstacles, vertical loading and inflation pressure. It is possible to obtain outputs from experiments on statical adhezor:

- Radial deformation characteristic (by vertical tire force loading);
- Torsion deformation characteristic (slip curve by twist moment);
- Size and shape of contact area and distribution of contact pressure;
- at following conditions:
- Loading (vertical);
- Tire pressure (under-inflation, overinflated tire, specified pressure);
- Size of radial deformation;
- Shape of obstacle etc.

5. Test of tire structure

Material parameters of long-fiber composite structural parts as the tire steel-belt are necessary input data for tire computational models (e.g. steel-cord belt) and for subsequent comparison of computational models with experiments. Knowledge is necessary of the behaviour of composites as belts under mechanical load. These data are obtained by experimental modelling of composite specimens and composite structural parts (matrixes and reinforcement) by static tensile, compression, shear and bending tests. The behaviour of such materials as tire belts under mechanical loading is in many ways different from the behaviour of commonly used technical materials such as steels. In composites, compared with metals, final mechanical properties can be controlled e.g. in the direction of the orientation of fibers-cords. Composites of tire also have elevated fatigue life, by one order higher material damping and are resistant to failure due to their ability to stop growth or decelerate propagation of cracks on the rubber matrix-cord interface. Tests of specific long-fiber composite materials with hyperelastic matrixes (namely steel.-cord belt test sample) are not standardized and neither are the shapes and dimensions of test samples, namely for tensile tests, which are for the observation of mechanical behaviour absolutely essential. For determination of material parameters of rubber matrixes and cord-reinforcements are necessary make experiments in agreement with standard specifications.

In composite samples or in samples with a certain content of composite layers of concern are the configurations of cords with respect to the direction of loading which results in a change of the stiffness characteristics. Therefore is necessary to design the geometrically parameters and shapes of one or multi-layer tested samples before experiments. The samples must have different:

- Angle of cord (with respect of the direction of loading – not only longitudinal and transverse orientated samples) – see figure 19;
- Material of cord (surface treatment);
- Form of cord (wire, thin wire);
- Number of layers (single-layer, two-layer – Figure 19, multi-layer);
- Specimen width, shape etc.

The author Krmela was designed multi-layer test samples with different wide *10, 15* and *25 mm* and of *length 120 mm*. The cord-angle orientations in single-layer specimens are *0°, 22.5°,*

45°, 67.5° and 90°. Two-layer specimens (Figure 19) are symmetrically orientated between top/bottom layer ±22.5°, ±67.5°, ±45° and asymmetrically orientated with cord-angles +0°/-45° and +67.5°/+22.5° (it is +22.5°/-112.5°, specimen D) with thickness 4 mm. Real single and two-layer specimens are presented Figure 20 as an example.

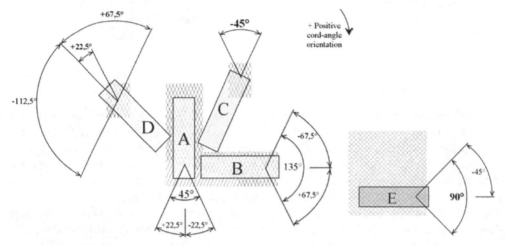

Figure 19. Two-layer specimens from plates with cord orientations 45° (left): A – lengthwise symmetrical specimen with 22.5°; B – transverse symmetrical specimen with 67.5°; C – asymmetrical specimen with +0°/-45°; D – asymmetrical specimen with +67.5°/+22.5°; Specimens from plates with cord orientations 90° (right): E – symmetrical specimen with 45°

Figure 20. Single-layer specimens of steel-cord belt with wire cord and two-layer specimens with thin-wire cord (right)

Also must be determined conditions for individual type of tests, namely:

- Statically tensile tests (uniaxial and biaxial);
- Statically bend tests;
- Statically tests of composites under combined loading states (combinations tensile with bend) - that are to be approximated tire real state during tire operational loading (predicate about real deformation behaviors of steel-cord belt plies);
- Corrosion tests in a corrosion chamber (exposition time);
- Dynamically test etc.

Statical tensile tests of steel-belt samples are important for obtaining knowledge about stiffness characteristics and material parameters. The conditions of the tensile tests are:

- Initial length between the jaws of the testing machine is *92 mm*.
- Elongation measured on the same length and also measured on *50* (or 25 mm) in centre of specimens.
- Rate of test is *10* or *25 mm/min*.

As output example from tensile test of two-layer belt for different cord-angle and cord-type Figures 21 and 22 give tensile force-elongation and stress-strain dependences (elongation measured on the length between the jaws).

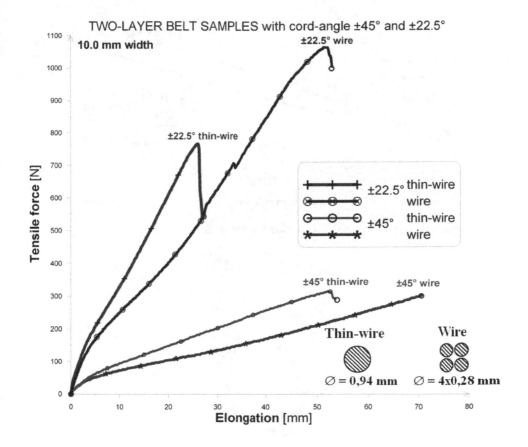

Figure 21. Outputs from tensile test of steel-cord belt samples – force-elongation dependences

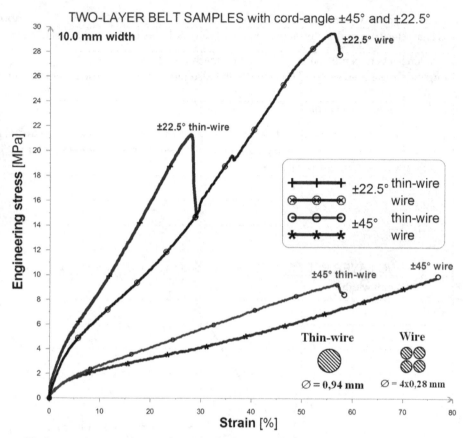

Figure 22. Outputs from tensile test of steel-cord belt samples – stress-strain dependences

Figure 23 presents some examples of specimens' failure after tensile test.

Figure 23. Failure of two-layer symmetrical ±22.5° and asymmetrical specimens +67.5°/+22.5° (right) after tensile test

The selected specimens were subject to statical tensile, also compression, shear and bending tests. Also testing conditions have been designed. Tests is necessary perform not only at ambient temperature 20°C but also at lowered and elevated temperature (from -30° into 180°Celsius).

6. Corrosion test of steel-cord belt

It will be possible uniform statically test conditions for samples affected by corrosion and samples without corrosion.

Selected single and two-layer composite test specimens are exposed to corrosion tests in a corrosion chamber *Gebr. Liebisch S 400 M TR* (for 500 or 265 hours in saline application by temperature at 70°Celsius – authors note: such extreme conditions should not ever appear in tire operations if proper conditions are kept) and to static tensile tests till the failure. The aim of these tests is to find the influence of the degrading process on the stiffness characteristics of the composite structures. Also will be investigated an influence of degree of degradation on the adhesive bond matrix-reinforcement.

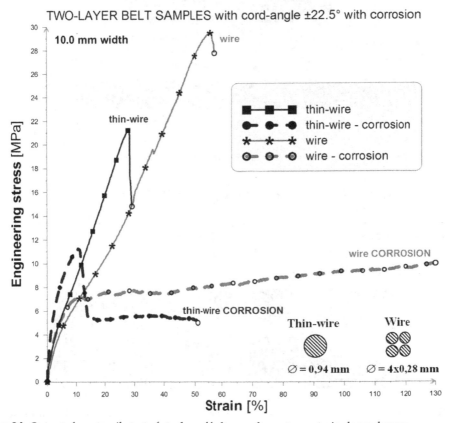

Figure 24. Outputs from tensile test of steel-cord belt samples – stress-strain dependences

The results obtained from tensile test will be compared with results obtained from tensile test on samples without corrosion. The experimental results of tensile tests of undamaged (non-corrosion) steel-cord belt ply (two-layer with cord-angle ±22.5°) for comparison analyses with belt ply after corrosion tests are shown in Figure 24 as dependences stress on strain.

The influence of corrosion on the stiffness and tensile force-elongation or engineering-stress dependences is sizable. The oxide film is strongly affected on failure of adhesive bond.

The fracture characters of test specimen after tensile and corrosion test in a corrosion chamber were accounted. The fracture of test specimen with 22.5° angle and thin wire cord is on Figure 25 as an example. The corrosion processes on cord surfaces is very dangerous. Therefore, it will be important to study also adhesive bond.

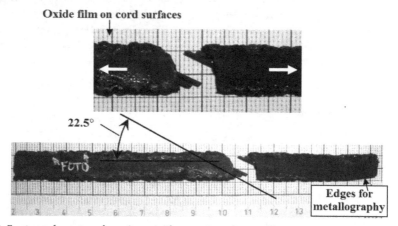

Figure 25. Fracture character of specimen with corrosion after tensile test

7. Metallography of Interface between Cord-Rubber

The light microscope is used for metallography observations of the adhesive bond between steel-cord and rubber after failure after corrosion test in corrosion chamber and statically tensile test and without corrosion too. The edges of steel-cord belt specimens were observed in detail – (see Figures 25 and 26).

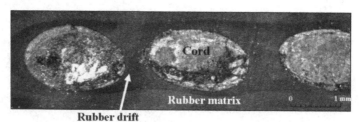

Figure 26. Interface between thin wire steel-cord/rubber drift/rubber matrix after corrosion and tensile tests

Microscopy with *100x-200x* zoom will be used for the evaluation of adhesive bonds from level of failure point of view. It appears that sufficient zoom from setting of failure level point of view (detection of delaminations, separations).It is necessary to prepare of samples for microscopy observation of structures so that the samples included different:

- Form of cord;
- Geometrical configurations and number of layers;
- Level of corrosion impact of steel-cords – adhesive bonds (without corrosion, easy corrosion, after corrosion test behind extremely conditions).

For selected cords were accounted:

- Uniformity of layer of rubber drift on cord surfaces;
- Surface treatment of cords;
- Interface between cord-rubber matrix after tire production;
- Structural change into microlocality of cord-rubber after corrosion.

The corrosion processes on cords are shown Figure 27. The results from microscopy observation are presented in Figures 28-33.

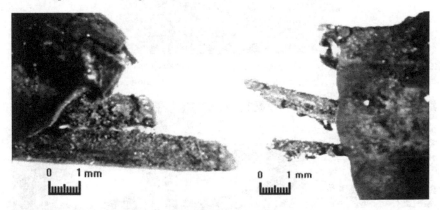

Figure 27. Corrosion processes on steel-cord surfaces – thin wire versus wire (right)

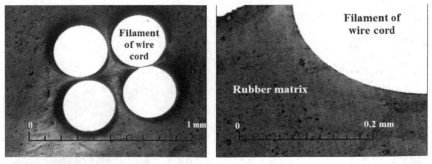

Figure 28. Good adhesive bond between wire steel cord 2+2×0.28 mm (cord consists of 4 filaments) and rubber after tire production (without corrosion)

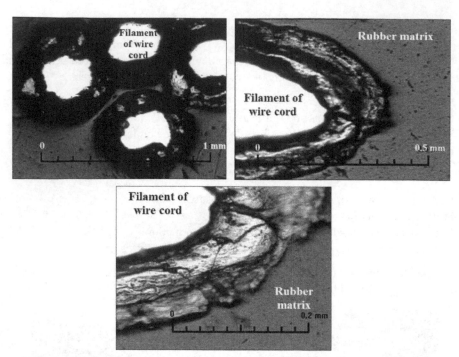

Figure 29. Damaged adhesive bond between wire steel cord 2+2×0.28 mm and rubber after corrosive attack (with extreme corrosion and tensile loading) with detail of oxide on filament surfaces

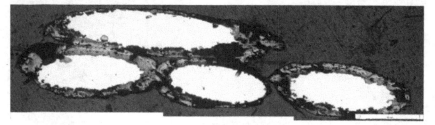

Figure 30. Damaged whole wire steel cord 2+2×0.28 mm

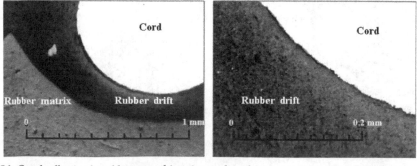

Figure 31. Good adhesive bond between thin-wire steel cord 0.94 mm and rubber drift after tire

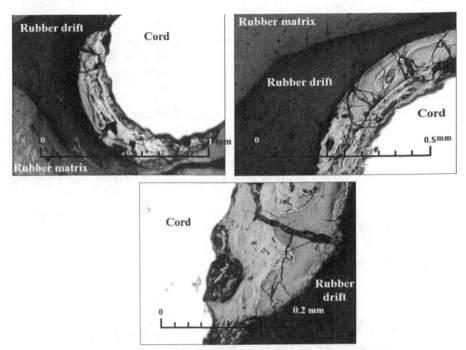

Figure 32. Damaged adhesive bond between thin-wire steel cord 0.94 mm and rubber drift after corrosive attack (with extreme corrosion and tensile loading) with detail of oxide on cord surfaces

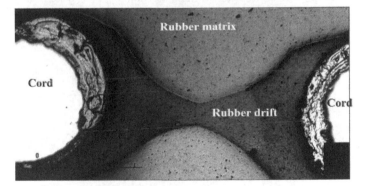

Figure 33. Damaged whole thin-wire steel cord 0.94 mm

On the base of corrosion tests is possible note:

- Arise of uniform surface corrosion;
- Fragility and hardness of corrosive layer;
- Quicker grow of oxides near cord surfaces;
- Fracture of oxide layer;
- Quality of cord-surfaces and surface treatment of cords with respect to corrosion attack;
- Decreased of material characteristics of steel-cord belt on the basic of tensile tests.

The adhesive bonds are influenced internal impacts (inserted during production, mounting) and external impacts (operating conditions, surrounding conditions etc.) or their interaction. It can be caused degradation on reinforcement-matrix adhesive bond, when its effect is failure into whole macro volume of tire which isn't permissible from safety aspect of vehicle.

- Any damage in the area of tire crown, namely into steel-cord belt plies, is perilous.
- If extreme corrosion on cords then cord surface treatment lost function of corrosive protection.
- If cords are with corrosion then adhesive bonds between cord-rubber matrix are damaged and safety of steel-cord belt plies and also tire is decreased.
- For predication of damaged belt ply is possible used combination of computational with experimental modeling.
- Corrosive attacks on reinforcing cords in whatever form can reduce the quality and operating safety of the whole tires.

8. Dynamic testing

In this part we continue with a description of dynamic materials behavior as well as a dynamic tire testing.

The tread displacement changes caused by the breaker angle changes were measured by an apparatus presented in Figure 34. The apparatus consists of a line laser, CCD camera and a computer with an appropriate measuring software. The CCD camera records the changes of the line laser spot, which copies the tire dimension changes. This system measures the main dimension of the rotating tire (the illuminated part of tread) at constant velocity. For more details see the work (Koštial et al 2006).

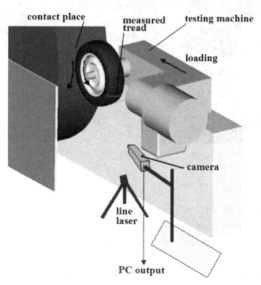

Figure 34. The measuring system for a tread deformation

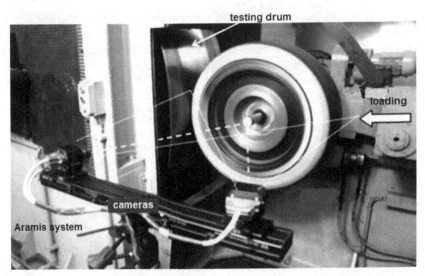

Figure 35. The measuring system for a sidewall displacement measurement

I. mixture		Properties	S	T	D
Hardness of vulcanizate [ShA]	73	Strength , LOP, [MPa]	19.23	17.08	23.27
Modulus 300% [MPa]	15	Elongation [%]	565	495	440
Strength [MPa]	16.2	Modulus 300% [MPa]	7.8	9.2	17.4
Elongation [%]	340	Hardness [ShA]	58	66	72
II. mixture		Elasticity [%]	35	18	48
Hardness of vulcanizate [ShA]	85	Tear resistance [kNm^{-1}]			
Modulus 300% [MPa]	12	at 20°C	57.1	37.5	74.8
Strength [MPa]	16.5	at 90°C	43.4	29.9	74.3
Elongation [%]	285				
III. mixture					
Hardness of vulcanizate [ShA]	60				
Modulus 300% [MPa]	7.2				
Strength [MPa]	17.9				
Elongation [%]	570				

Table 1. The physical parameters of bead core blends (left) and other part of the tire (right part).

The sidewall displacement changes caused by both, bead core and the breaker angle, were measured by a contactless system Aramis. This system is able to measure changes of the displacements (radial and axial) during the rotation of the tire. More experimental details about the apparatus are in the work (Koštial et al 2005) (see Figure 35.) The statistically evaluated precision of both described equipments at the actual arrangement of the apparatus was 0.05mm (the result of ten independent measurements on the same tire). The radial loading was *7360 N* (that is 80% of the maximum available load) with a tire inflation of *290 kPa*. After the tire conditioning (30 min at the velocity 80 km/h) the tire inflation increased (due to heat generation at the tire movement) to a pressure of *310-320 kPa*. The considered testing velocities were *10, 50, 80, 120, 150 and 180 km/h*. The reference velocity was *10 km/h* (zeroth stage). The image of a sidewall obtained at *10 km/h* is compared with other images obtained at different velocities. The difference between those images determines a displacement.

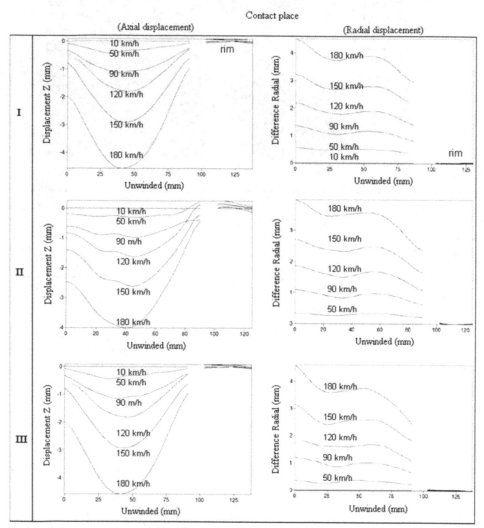

Figure 36. The axial displacement changes at the contact (left side) and radial displacement (right side) for three different bead core blends. Unwinded means a length of a given sidewall part.

The physical parameters of the bead core blends are described in the left part of Table 1. The characteristic physical parameters blends used for the construction of the other part of a tire are collected in Table 1 – right part. The abbreviations used in Table 1 mean: *S* – sidewall rubber mixture, *T* – tread rubber mixture and *D* – depositional rubber mixture on the steel cord.

The measurements of the sidewall displacement at the contact of the driving drum and tire presented in Figure 36 also show the smallest radial and axial value for the sidewall containing the blend *II*. The upper line is for the reference velocity *10 km/h*. The minimum showing at the bottom of every picture (for axial displacement) corresponds to the velocity of *180 km/h*. According to the presented results it is possible to conclude that in the current case the blend *II* has the best properties (the smallest displacement of a sidewall means higher mechanical stiffness and smaller rolling resistance, for instance) for the sidewall construction. Further we will analyze the changes of the tread and sidewall displacement caused by changes of breaker angle with unchanged bead core blend. The tread and sidewall materials were also the same according to the description presented above (Table 1).

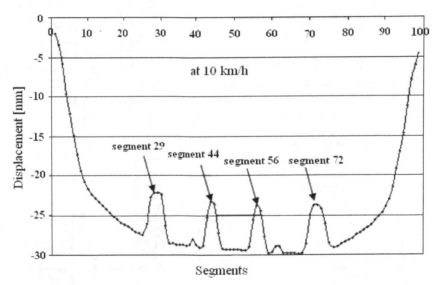

Figure 37. The electronic picture of a tread with characteristic points marked as "segments"

Figure 37 shows the electronic picture of the tread with characteristic points marked as "segments" obtained by an optical system with a line laser. The tread displacement changes in radial direction (at above defined velocities) caused by a breaker angle change are visible in Figure 38.

Rising of the breaker angle changes the shape of the tread deformation from convex to concave. The best solution was obtained for the breaker angle equals *27°*, where practically the full profile of the tread is in contact with the road. In order to study the influence of breaker angle changes on "sidewall displacement dynamic" at different velocities we also tested the sidewall displacement changes (measured by ARAMIS, breaking angle equal to

20° and 27°). The experimental results of both the axial and radial displacement in this case were compared with those obtained by the *FEM* simulation in *ABAQUS* environment. It is possible to see a good agreement between both simulated and measured curves. Differences occur at higher velocities for radial displacement. Both, the experimental and simulated axial displacements for different velocities and chosen breaker angles (*20°* and 27°) are displayed in Figures 39 and 40. The corresponding simulation and experimental results obtained for radial displacement also show the best results (the highest axial displacement) for breaker angle 27°.

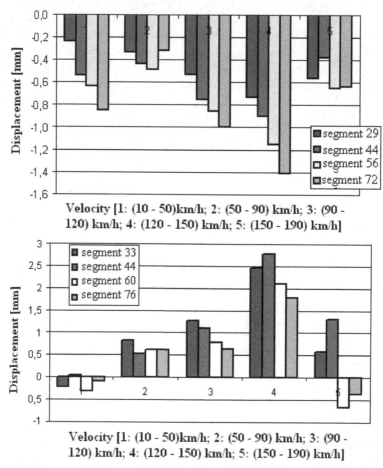

Figure 38. The displacement differences for the breaker angle 20° (variant 1 - left) and the displacement differences for the breaker angle 27° (variant 4 - right)

On the basis of these results it is possible to state that the higher the breaker angle the higher the displacement is in both axial and radial directions. In other words, the tire "grows" with rising of the breaker angle.). These results support the highest values of elongation and a relatively high value of strength and elasticity which provide also the so called driving comfort. On the other hand, the highest tear resistivity, 300 % modulus and hardness of the depositional rubber mixture on a steel cord provide all together good tire safety (see parameters in Table 1).

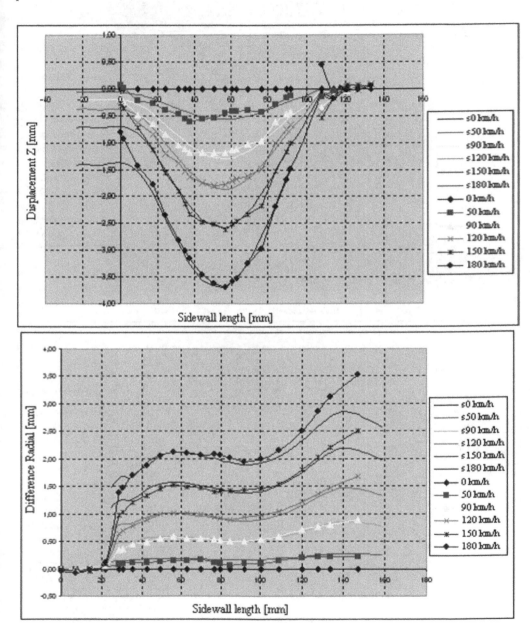

Figure 39. The axial (left) and radial (right) displacement of the sidewall for 20° breaker angle (s-simulation)

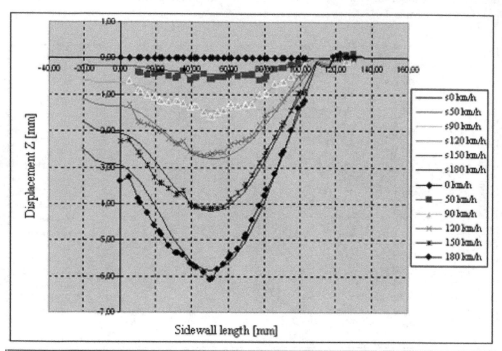

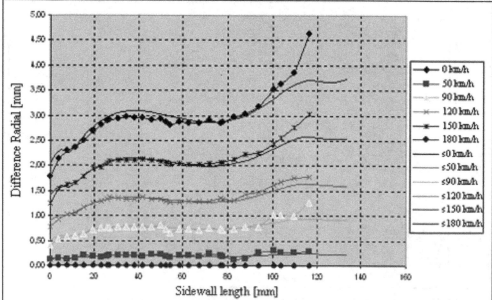

Figure 40. The axial displacement of the sidewall for 27° breaker angle (s-simulation)

9. Conclusion

The chapter presents the large scale view on the problem of tire safety concluding materials aspects, damage and experimental testing of tires and tire components materials. For more information about the solution of tyres, see reference (Krmela 2008).

Author details

Pavel Koštial and Ivan Ružiak
VŠB – Technical University of Ostrava, Faculty of Metalurgy and Material Engineering,
Department of Material Engineering, Ostrava, Czech Republic

Jan Krmela
University of Pardubice, Jan Perner Transport Faculty,
Department of Transport Means and Diagnostics, Pardubice, Czech Republic

Karel Frydrýšek
ING-PAED IGIP, VŠB – Technical University of Ostrava, Faculty of Mechanical Engineering,
Department of Mechanics of Materials, Ostrava, Czech Republic

Acknowledgement

The work has been supported by the Czech grant projects MPO FR-TI3/818 (sponsored by the Ministry of Industry and Trade of the Czech Republic) and by the Slovak-Czech grant project 7AMB12SK126 (sponsored by the Ministry of Education, Youth and Sports of the Czech Republic).

10. References

Ferry, J. D. (1980). *Viscoelastic properties of Polymers*; John Wiley&Sons, New York, 1980

Jakubíková, Z.; Skalková, P.; Mošková, Z. (2007). Mechanical, Thermal Characterization of low Density Polyethylene (LDPE)/Carboxymethylstarch (CMS) Blends, In: *1st Bratislava Young Polymer Scientists workshop BYPoS*. 2007, Bratislava, Slovakia

Jančíková, Z. (2006). *Umělé neuronové sítě v materiálovém inženýrství*, GEPARTS Ostrava, Czech Republic, 2006, written in Czech language

Jančíková, Z., Švec, P. (2007). *Acta Metallurgica Slovana*, 2007, 13, 5

Koštial, P., Mokryšová, M., Klabník, M., Žiačik, P.; Kopal, I.; Hutyra, J. (2006). *Rubber World.* 2006, 233, 4, 18-20

Koštial, P., Mokryšová, M., Kopal, I.; Žiačik, P., Rusnáková, S., Klabník, M. (2005). 12th International metrology congress. 2005, Lyon, Collége Francais de Métrologie, France

Krmela, J. (2008). *Systems Approach to the Computational Modelling of Tyres - I. Part*, Czech Republic, 2008, pp.1-102, ISBN 978-80-7399-365-8, book written in Czech language

Maier, P., Gand Göritz, D. (1996). *Kautschuk Gummi Kunststoffe*. 1996, 49, 18
    http://ao4.ee.tut.fi/pdlri

Medalia, A. I. (1978) *Rubber Chem. Technol.*, 1978, 51, 437

Murayama, T. (1978). Elsevier scientific publishing company Amsterdam-Oxford-New York, 1978

Payne, A. R. (1965), Wiley Interscience, New York, 1965

Schaefer, R. J. (1994). *Rubber World*. 1994, 11, 17

Sepe, M. P. (1998). *Plastics Design Library Norwich*, New York, USA, 1998

Simek, I. (1987). *Fyzika polymérov*; SVST, Bratislava, Czechoslovakia, 1987, written in Slovak language

Wang, M. J. (1998). *Rubber Chem. and Technol.*, 1998, 71, 3

The Lightweight Composite Structure and Mechanical Properties of the Beetle Forewing

Jinxiang Chen, Qing-Qing Ni and Juan Xie

Additional information is available at the end of the chapter

1. Introduction

The development of lightweight, energy-saving structures is believed to be one key solution to current world problems, such as increasing population, resource shortages, and environmental pollution. Acquiring inspiration from living creatures is an effective approach because these forms have evolved over millions of years to adapt to the natural environment [Thompson, 1945]. Examples of evolutionary adaptations provide evidence of how nature has helped organisms to overcome their structural weaknesses [Thompson, 1945; Wainwright et al., 1976; Hepburn, 1976]. Motivated by our insight that the forewings of beetles should have both high strength and minimal weight, which are needed both for defense and for flight, the first author of this paper has been studying their architecture since 1997 [Chen et al., 2000; Chen et al., 2001a; Chen et al., 2007a; Chen et al., 2007b]. However, beetle forewings were considered nonliving materials in early work [Ishii, 1982], which may be the reason why research on beetle forewings and their biomimetic applications has been limited to two-dimensional fiber orientations [Gullan & Cranstion, 1994; Zelazny & Neville, 1972]. To date, we have comprehensively investigated their three-dimensional structures and obtained new information on their biological morphology [Chen et al., 2007b; Chen et al., 2001a; Chen et al, 2002], and we have gained new insight into their mechanical properties [Chen et al., 2000; Chen et al., 2007a; Ni et al., 2001]. Our investigations have not only confirmed the above insights but also led us to discover a new type of lightweight biomimetic composite that is more complicated and more delicate than the presently used honeycomb structure. This new composite features a completely integrated honeycomb structure with fiber-reinforced trabeculae at the corners of the honeycomb cores [Chen et al., 2005; Chen et al., 2012].

In this chapter, we present insight into the lightweight composite structure of the beetle forewing, including its mechanical properties and practical applications.

2. The lightweight composite structure of the beetle forewing

2.1. Experiment

Two species of adult beetles, *A. dichotoma* and *P. inclinatus*, were used as experimental samples and are shown in Fig. 1(a) and (b), respectively.

Since living samples were used, their careful treatment by specialized methods to ensure that their structures would remain unchanged during the sample preparation process was highly important. Here, three specialized treatments were used for comparison of the forewing structural damage caused by removal from the living beetle body: (1) living samples (tests are conducted within 1-3 minutes); (2) traditional mistreatment; and (3) the living samples left at room temperature for one day or, in some cases, even several months. The results show that by using method (3), which may result in drying due to loss of water and partial damage of the cell membrane, the original structural integrity of the forewing still remains [Chen et al., 2002]. In this experiment, unless otherwise noted, method (3) was used for sample preparation.

The observation positions were set using Cartesian coordinates, and the names of the different regions of the forewing are shown in Fig. 1(c). The cross sections were prepared by slicing the forewing along the X or Y line, after which the inner structure of the forewing was observed by removing a portion of the lower lamination with forceps.

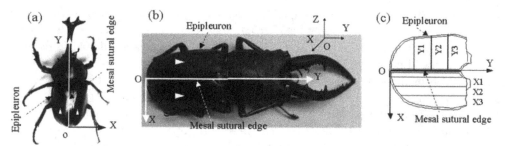

Figure 1. Two species of adult beetle and the Cartesian coordinates (△, the fore-wings). (a) male *A. dichotoma*, (b) male *P. inclinatus*, (c) observation sites.

A strong alkali can be used to examine the composition of the forewing; the chitin fibers do not readily dissolve in the alkaline solution, while the proteins dissolve easily. Once the protein matter has been removed by the alkali solution, the structures of the trabeculae can be readily observed. Therefore, some samples were prepared by boiling in a 10% KOH solution [Zelazny & Neville, 1972] for three hours. This treatment does not affect the orientation of the chitin fibers.

The main experimental instruments used in this study were an environmental scanning electron microscope (Nikon ESEM-2700 at 20 kV and 460 Pa) and a general SEM (Hitachi, S-510 at 15 kV).

2.2. Results and discussion

2.2.1. General structure of the beetle forewing

The forewing photographs of *A. dichotoma* are presented in Fig. 2(a-e). Fig. 2 (a, c) and Fig. 2 (b, d, e) contain images captured with a camera under a penetrating light with a short focus

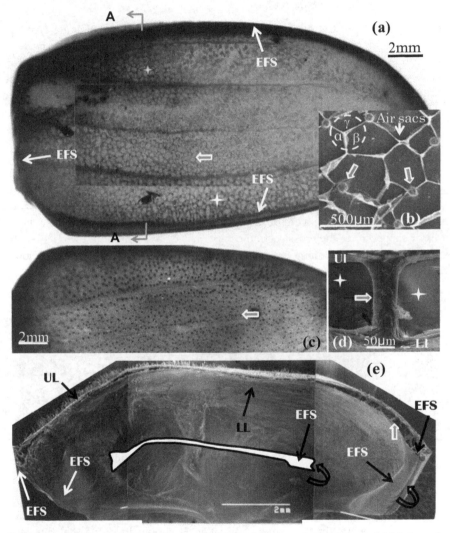

Figure 2. The integrated structure and edge sealing of the forewing of *A. dichotoma*. (a) Meshwork and trabeculae (dots) in a fresh forewing, (b) the internal structure and the lower lamination were eliminated by forceps, (c) distribution of trabeculae (dots) in the forewing (following 10% KOH treatment), (d) a trabecula, (e) the cross sectional view of the A-A section, and the middle part of the figure is a simple sketch of the cross section. EFS: edge frame structure, asterisk: air sacs, thick arrows: trabeculae, round arrows: concavo-convex junction, Ul: upper lamination, Ll: lower lamination.

time and with a SEM, respectively. The many lines Fig. 2(a) apparent in those photos are actually air sacs or tracheae [Chen et al., 2002]. The images in Fig. 2 show the samples after three hours of boiling in the 10% KOH solution. The air sacs and the tracheae that can be observed in Fig. 2(a) have disappeared here because they were dissolved during the KOH treatment. However, many black solid dots (Fig. 2c) are visible, each of which represents an individual trabecula [Chen et al., 2002].

As shown in Fig. 2 (e), the entire beetle forewing is composed of a single type of frame structure. Fig. 2 (e) shows an electron micrograph of this frame structure at an X-Z plane cross section (see Fig. 1). It was found that the forewing cross section consists of three subdivisions; the upper lamination, the lower lamination and the central void, with the large voids located at both the mesal sutural and the epipleuron edges (see Fig. 1). The middle portion of the forewing is a sandwich structure containing a wide void area.

2.2.2. Trabecular structure in the beetle forewing

The longitudinal shapes of typical forewing trabeculae in *A. dichotoma* are shown in Fig. 3 (a, b), and a cross section is shown in Fig. 3(c). The trabecular size and shape can change according to location in the forewing. Both trabeculae with a straight shape (Fig. 3a) and with a curved shape (Fig. 3b) were observed, although the majority of the trabeculae were of the straight shape kind. The cross section of any trabecula is similar to the circular structure presented in Fig. 3(c). The diameter of a trabecula is larger near the locations of joints with the upper and lower laminations and is smaller in the void region between.

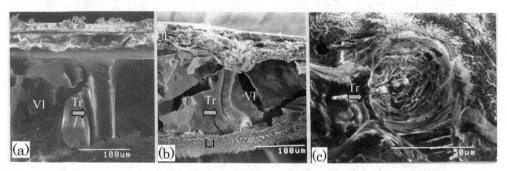

Figure 3. The trabecular (Tr) shapes. (a) Straight cylinder, (b) curve cylinder, (c) section view. Ul, upper lamination, Vl, void lamination, Ll: lower lamination.

Fig. 4 shows longitudinal sections of the trabeculae, which allow for clear observation of the chitin fiber arrangement within the trabeculae. In Fig. 4(a), the chitin fibers Fb (or marked ☆) form a distinct layer within the upper lamination. It should be noted that the fiber angle within this layer is nearly perpendicular (at 90 degrees) to the paper and that the neighboring chitin fiber layers (Fl or ↘) are instead horizontally arranged and connected with the trabecula in a curving manner. This connection between the chitin fiber layers in the upper lamination and the trabecula can also be seen clearly in Fig. 4(b). Fig. 5 shows micrographs of trabeculae after having been boiled in the 10% KOH solution for three hours.

Due to removal of the protein matter, the chitin fiber arrangement on the outside of the trabeculae can easily be observed. The chitin fibers in the left image (Fig. 5a) are in a straight arrangement, while the fibers in the right image (Fig. 5b) have a spiral arrangement on the trabecular surface. The central regions of the trabeculae become hollow (Fig. 5c) due to dissolution of the internal proteins during the KOH solution treatment. Thus, the central portion of the trabecular shaft is composed mainly of protein [Chen et al., 2001b].

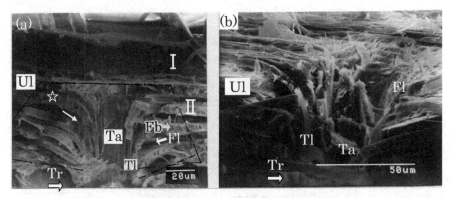

Figure 4. The trabecular structure in the forewing of *A. dichotoma*. (a) longitudinal section with the trabecular root marked by the lines; (b) a naturally rupture trabecula. Ul, upper lamination. Fl, fibers from left to right. Fb, fibers from front to back. Ta, trabecular central part. Tl, trabecular cylindrical layers. Tr, trabecula.

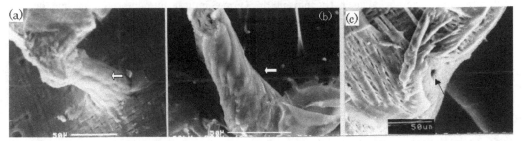

Figure 5. Trabecular structures treated with KOH, (a) the straight orientation (b) the spiral curved orientation of the chitin fiber at the outside, (c) the hollow (↖) at the center of the trabecular root.

2.2.3. Schematic model of the forewing structure

Based on these observations and analyses of the beetle forewing and its inner trabecular structures, it is evident that the general forewing structure in *A. dichotoma* is a sandwich frame structure with a central void region. However, the details of the forewing microstructure are extremely complex. These models for the forewing and trabecula structures are proposed to assist in the development of biomimetic composite structures. According to our observations, we have developed a sandwich plate structure model with a central void layer and trabecular struts [Chen et al., 2005]. It is well known that sandwich plates with a central void layer are typical of lightweight structures and already have been applied in many fields [Jung & Aref,

2005; Matsunaga, 2002; Nguyen et al., 2005; Mania, 2005; Hosur et al., 2005]. The presence of the sandwich plate structures with trabeculae in the beetle forewings indicate that the beetle wings have been evolutionarily optimized for competition in nature.

Fig. 6 shows the sandwich plate model that was inspired by the forewings of the beetles. The plate consists of upper and lower lamination layers and a core layer with trabeculae and honeycomb walls (Fig. 6a). There are many trabeculae positioned throughout the honeycomb, and they are located at the corners of the honeycomb units (Fig. 6b).

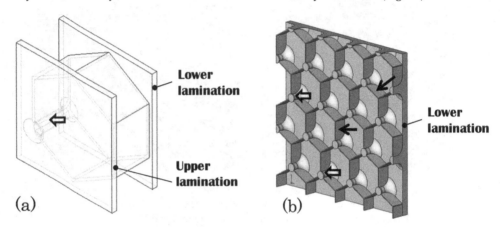

(a)   (b)

Figure 6. Three-dimensional sketch of a trabecula-honeycomb cell. (a) A cell. (b) the assembly of cells without an upper lamination layer. The thin and thick arrows indicate the honeycomb wall and the trabeculae, respectively.

2.2.4. *Schematic model of the trabecular structure*

Based on our observations, a schematic structural model of the trabeculae in *A. dichotoma* was generated and is shown in Fig. 7. This model indicates the following: (1) the substance in the central part of the trabecula is protein (see thin arrow in Fig. 7a); (2) The upper and lower laminations are comprised of chitin fiber layers that either extend into the trabeculae continuously in a curved shape (see Fl in Fig. 4a and Fig. 7) or connect to the trabeculae by surrounding the circumference of the trabecular shaft in a spiral fashion (see Fb in Fig. 4a and Fig. 7). Through peeling tests, we have confirmed that, when comparing the trabeculae of the upper and lower laminations, the peeling resistance of trabeculae in a general region is approximately three times greater and in a local region is approximately thirty times greater [Ni et al., 2001]. Thus, the use of the trabecular structures to join the upper and lower laminations effectively increases the interlaminar strength of the composite structures.

2.3. Conclusions

To develop lightweight biomimetic composite structures, the forewing structures of *A. dichotoma*is were investigated. Structural models for both the beetle forewings and the forewing trabeculae were also established, and the results obtained are as follows:

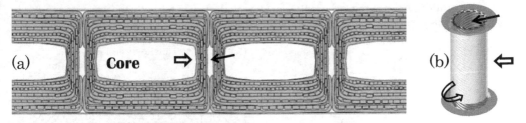

Figure 7. Schematic view of reinforcing fibers. a, Cross section of a forewing. b, Three-dimensional view of a trabecula. The thin and thick straight arrows show a void (protein) inside a trabecula and itself, respectively. The circular arrow shows a spiral fiber surrounding the trabecula (the lower part shows the structure of a trabecula without a protein).

1. It was found that both species of beetle forewings have a sandwich plate structure and a frame structure with a central void layer and interconnecting trabeculae. These characteristics are typical of lightweight composite structures.
2. The chitin fibers in the trabeculae connect with a curved shape continuously to the chitin fibers on the upper and lower laminations, and it is clear that this structure greatly increases the peeling resistance of the laminated composite structures.

3. Interlaminar reinforcement mechanism of the beetle forewing

The fracture toughness of fiber-reinforced composite materials used in engineering applications mainly depends on the interlaminar and/or intralaminar fracture toughness values, with the interlaminar and/or intralaminar interface tending to be the critical location of weakness. Investigations of the interfacial behavior have been undertaken by many researchers, such as evaluation of the interlaminar and/or intralaminar fracture toughness [Iwamoto et al., 1999a], the interlaminar delamination resistance [Turss et al. 1997], and the role of bridging fibers in crack propagation [Sekine & Kamiya, 1987; He & Cox, 1998]. In recent years, 3D textile-reinforced composites have been developed toward improving the interlaminar and/or intralaminar fracture toughness values. On the other hand, many examples of composite structures already exist in nature, and their structures have already been optimized by evolutionary pressures over the course of a deep ecosystem history.

This section focuses on the mechanical properties and the three-dimensional interlaminar reinforcement mechanism of the *A. dichotoma* beetle forewing.

3.1. Experiment

The specimen utilized was the living forewing of the adult male *A. dichotoma* beetle (Fig. 1a, Δ arrowhead). The width of the fresh specimen was 2.1 mm. A tensile device (SHIMAZDU, AUTOGRAPH DSC-10T) with a custom low load cell was used for the interlaminar peeling test, which was conducted with zipper forceps, as developed by the authors. The tensile loading direction was vertical to the peeling surface, as shown in Fig. 8(a). To prevent sliding of the specimen at the zippers, the inner surface of the forceps was adhered to the specimen with glue tape. The peeling schematic is indicated in Fig. 8(b). In the present study, 10 specimens were tested at a tensile speed of 5 mm/min and a peeling distance of 7 mm.

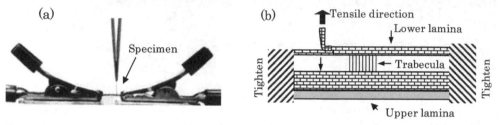

Figure 8. A peeling experiment; (a) peeling device, (b) a peeling schematic.

3.2. Results and discussion

3.2.1. Peeling characteristics of the forewing of A. dichotoma

Micro Fractograph of Peeling Surface Fig. 9 presents a micrograph of a forewing after the peeling test by illuminating the specimen from behind with penetrating light, which is a new technique developed for observation of the forewing structure. Many black points and ducts can be observed in this image, where the black points can be interpreted as trabeculae and the ducts, as tracheas. Fig. 10 (a) is a micrograph taken at the Δ arrowhead mark in Fig. 9 and shows the fracture characteristics of the first trabecula during the peeling process. The peeling test was started by applying pulling force to the lower lamina of the forewing. When the delamination crack tip met the first trabecula (Fig. 10a, Δ arrowhead), a portion of the lower lamina tore, while the remaining portion was held in tension. In this manner, the peeling tip continued to develop and propagate between the chitin fiber laminae. A fracture micrograph illustrates a peeling surface in the region of the ▲ arrowhead marked in Fig. 9 and has been shown in Fig. 10(b). The resulting fracture patterns on the peeling surface had an appearance similar to that of a tree root (Fig. 10b, Δ arrowheads) with details of this phenomenon presented in Fig. 11(b) and Fig. 13.

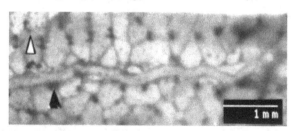

Figure 9. A photograph of a fore-wing after peeling test and black points are trabeculae.

Fig. 11(a) illustrates the fracture pattern of the peeling surface without any trabeculae present. Although the chitin fibers are arranged in rows (Δ arrowhead), some of have torn (▲ arrowhead) during the peeling process. Fig. 11(b) shows a fracture micrograph of the chitin fibers at a trabecular root. It was that the chitin fibers (Δ arrowhead) formed very thin

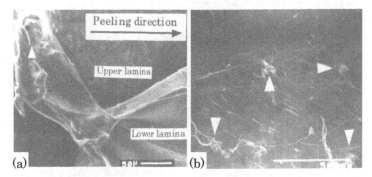

Figure 10. Fractographs of a forewing; (a) fracture of first trabecula, △: trabecula, (b) a peeled chitin fiber lamina, △: roots of the trabeculae.

layers here, with Fig. 11(c) illustrating a side view of the peeling process. When the peeling tip had propagated about half of the thickness of the lower lamina, a few bridging fibers can be seen between the lower and upper peeling surfaces.

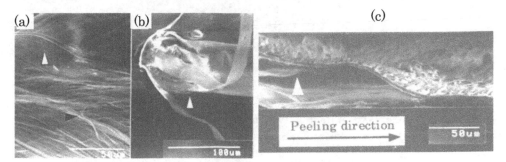

Figure 11. Fractographs of a forewing; (a) peeled surface without a trabecula, (b) trabecula root and thin chitin fiber layers, (c) a peeling process from a side view. △: bridging fibers.

3.2.2. Load peaks and trabeculae in a peeling test

A representative load-displacement curve, with load F and displacement δ, is presented for a forewing peeling test in Fig. 12(a). The force, $F(\delta)$, was proportional to the resistance force because the tensile direction was held vertical to the peeling surface, as described above. Here, the F-δ curve is named the "peeling test" and the curve is given by $F(\delta)$. Many individual load peaks were present in the peeling curve, and their shapes were similar to the standard load peak shape, although their peak values were quite different. In the peeling test, the minimum value of each ravine was not equal to zero but instead held a nearly constant value during the peeling process. Here, the jth values of the load peak and the ravine minimum are indicated by $F_{j1}(\delta)$ and $F_{j0}(\delta)$, respectively.

The number of individual load peaks in the F-δ peeling test was compared to the number of trabecular roots, which were observed by SEM, and these values were found to be nearly equal [Ni et al., 2001]. Accordingly, it was determined that each trabecula contributed a single load peak in the peeling curve. In other words, the load peaks resulted from breakage of the individual trabeculae during the peeling process.

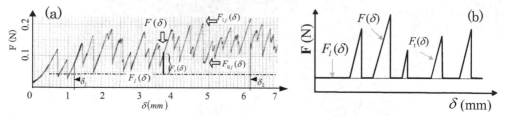

Figure 12. A representative load-displacement curve (a) and its model (b) in a forewing peeling test.

Peeling Curve As mentioned above, the load peaks in the F-δ peeling curve were attributed to specific contributions from the individual trabeculae. Based on this fact, the peeling curve, $F(\delta)$, was considered to be comprised of two components, i.e., $F_l(\delta)$ (Fig. 12a) due to interlaminar delamination of the chitin fiber laminae and $F_t(\delta)$ due to trabecula failure. Thus, $F(\delta)$ may be represented by equation (1):

$$F(\delta) = F_l(\delta) + F_t(\delta) \tag{1}$$

$F_l(\delta)$ depends mainly on the properties of the chitin fibers, the protein matter and fiber adhesion. If the interlaminar debonding force, $F_l(\delta)$, was constant during the peeling process and the trabeculae were distributed along only one line in the propagation direction, then a schematic of the peeling curve can be represented as Fig. 12(b). In other words, each load peak, $F_t(\delta)$, results from the fracture of one trabecula, and the line, $F_l(\delta)$, at the ravine minimum results from the delamination resistance between chitin fiber layers without the trabeculae.

However, the horizontal line, $F_l(\delta)$, cannot be observed in the F-δ peeling curve as indicated in Fig. 12. This is attributed to the trabeculae being distributed intermittently throughout the plane (Fig. 11c) with the bottom of a ravine, $F_{j0}(\delta)$, generated from the delamination of laminar fibers that are piled upon each other. In comparing Fig. 12 (b) with Fig. 12 (b), it was not possible to fully distinguish $F_l(\delta)$ and $F_t(\delta)$. However, the value of $F_l(\delta)$ should be less than or close to the minimum of all ravine bottoms, F_{min0}, i.e.,

$$F_l(\delta) \leq F_{min0} = Min\{F_{j0}(\delta)\} \tag{2}$$

where F_{min0} is the maximum of $F_l(\delta)$ in the following calculation.

3.2.3. 3D Reinforcement mechanism of the beetle forewing

Fracture Type of Trabecular Roots Three representative fracture modes of the trabecular roots were observed during the forewing peeling test (Fig. 13). Fig. 13(a) presents type A, which possesses a convex shape, and also shows a schematic of the longitudinal section shown in the fractograph. In type A, trabecular roots in the lower laminae remained following peeling such that the setae (▲ arrowhead) could be observed. In type C, a smooth surface (Fig. 13c) has had the trabecular roots in the lower laminae completely removed. In contrast to this, type B was between type A and type C (Fig. 13b). Although three types of failure shape for trabecular roots were observed, the load peaks in the peeling curve were similar to each other. In the following section, possible reasons for this are considered.

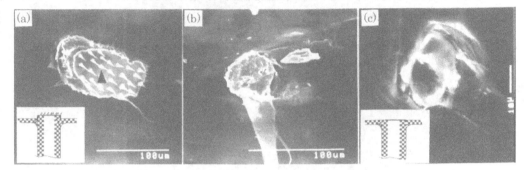

Figure 13. Fractographs of three typical trabecular root.

According to the trabecular structure, the chitin fibers within the trabeculae are thin, continuous, and curved (Fig. 14, ▲ arrowhead) and interconnected with each other between the chitin fiber laminae. During the peeling process, the peeling tip continues to propagate horizontally (see Fig. 11c) in the absence of a trabecula. When the peeling tip encounters a trabecula, however, its progress is impeded by the chitin fibers of the trabecular; meanwhile, the peeling tip around the outside of the trabecular root continues to propagate, which results in a stress concentration at the trabecular root itself. That is, the chitin fibers of the trabecula initially withstand the higher stress and then become stretched into a straight line until the trabecula root fails (Fig. 14a). Therefore, the chitin fibers within the trabecular root

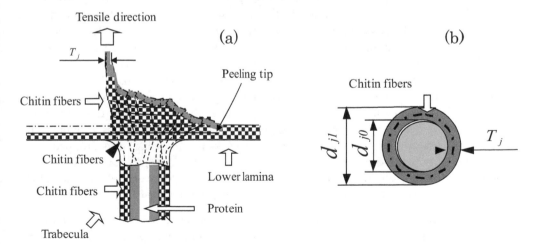

Figure 14. A chitin fiber reinforcement model (a) and a cross sectional trabecular model (b).

fracture in tension. Thus, the chitin fibers in the trabeculae have a high resistance to the applied load, and their failure results in the load peaks noted in the F-δ peeling curves as shown in Fig. 12. This also results in similar peeling curve peak shapes for the three fracture types (Fig. 13).

Failure Strength of Chitin Fibers in the trabecular root The failure strength of chitin fibers in the trabecular root has been discussed using the proposed models (Fig. 14). Using equation (1), the load peaks, F_{j1} (j =1, 2,..., n, the number of trabeculae), and the resistance force, $F_l(\delta)$, of the interlaminar delamination in the peeling curve, the average failure strength, $\overline{\sigma}_b$, of the chitin fibers in the trabeculae was calculated as follows:

$$\overline{\sigma}_b = \frac{\sum_{j=1}^{n}\left[F_{j1} - F_l(\delta)\right]\Big/n}{\sum_{j=1}^{n}S_j\Big/n} \tag{3}$$

where S_j is the failure area of the chitin fibers and was calculated using the thin ring area of chitin fiber layers within the trabecular root (Fig. 14b). The thickness of the thin ring was considered to be equal to the chitin fiber lamina thickness, T_j, and S_j was given by equation (4) ·

$$S_j = T_j\left(d_{j1} - T_j\right)\pi \tag{4}$$

The justification for using the cross section of the thin ring area is given as follows: first, the trabecula consists of two sections, which include the central part that is mainly protein and the outside that is chitin fiber laminae. This result was confirmed by treating the forewing with 10% KOH at 100°C for 3 hours [Iwamoto et al., 1999b]. After treatment, an empty space was observed in the central region of the trabecula (see Fig. 5c, arrowhead part), which was attributed to dissolution of the protein in the KOH solution. Therefore, it is clear that the central region of each trabecula contains no chitin fibers and is instead mainly composed of protein. Second, considering the fracture of chitin fibers during the peeling process (see Fig. 14a), when the peeling tip of the chitin fibers progresses around the outside of the trabecular root, the chitin fibers at the remaining intact side bear a higher applied load than those in the delaminated side. However, in the cross section of the trabecular root thin ring, the stress is uniformly accommodated; in particular, this is true after the outside chitin fiber laminae of the trabecula are peeled apart (delaminated) because the chitin fiber layer is very thin. This hypothesis was supported by the fact that the load peaks in the peeling curve dropped when a trabecular root fractured during the peeling process (Fig. 12). Third, the trabecular diameter in the forewing of *A. dichotoma* is very small and only on the order of tens of microns. In light of the above reasons, and for simplicity, equation (4) was thus adopted.

Moreover, for equation (3), only data collected in the region between 1 mm and 6 mm of displacement were used because the peeling front had not always progressed through the full specimen width in the initial and final failure regions. Thirty load peaks in the region of

1 mm to 6 mm were used for the calculation, although there were 44 load peaks in the total peeling curve as shown in Fig. 12. As a result, the trabecular root failure load and the resistance force, $F_l(\delta)$, of interlaminar delamination were found to be 0.156±0.029 N and 0.049 N, respectively. The thickness of the peeled chitin fiber laminae, T_j, was 4.5 µm, which remained nearly constant in the entire peeling region from 1 mm to 6 mm.

Next, the external diameter of the peeled trabecular root, d_{j1}, was determined. In practice, the cross section of the trabecular root is not circular but instead is closer in shape to an ellipse. Thus, the average of the long and short axes of the ellipse was used as d_{j1}. The results of d_{j1} as measured by SEM indicated the average external diameter of 30 trabecular roots was 55 µm · Using all of the results obtained above, the failure strength, σ_b, of the chitin fibers within the trabecular root was calculated as approximately 149 MPa. This result is close to the tensile strength of 147 MPa in the same 0° direction of the chitin fibers. Therefore, during the peeling test, the failure strength of the chitin fibers within the trabecular root ultimately resulted from tensile fracture of the chitin fibers, as mentioned above.

Interlaminar Reinforcement Mechanism The chitin fibers of the forewing trabecular root were found to have the following characteristics: a curved and continuous shape; bonded to both chitin fiber laminae and to each other; with the angle of about 90° between the lower laminae and the trabecula. The increment of interlaminar strength is due to the tensile failure of the chitin fibers within the trabecular root, as determined by the peeling process. These results indicate a creative approach for researchers to overcome the interfacial weakness issues currently associated with laminated composite structures.

Strengthening Effects due to the Trabeculae The strengthening effects due to the forewing trabeculae were investigated. Here, the local strengthening effect, λ_l, is calculated by equation (5), i.e., λ_l is the ratio of the average failure load of the chitin fibers in the trabecular root due to the resistance force of the interlaminar lamination, $F_l(\delta)$, in units of inverse length.

$$\lambda_l = \overline{\sigma_b} \cdot T_j \left/ \frac{F_l(\delta)}{B} \right. \cong 29 \tag{5}$$

Using the data presented in Fig. 12, the average failure load of the chitin fibers in the trabecular root, $F_t(\delta)$, the resistance force of the interlaminar lamination, $F_l(\delta)$, and the specimen width of B = 2.1 mm were all calculated. The local strengthening effect, λ_l, was found to have a value of approximately 30. Therefore, the 3D strengthening effect of the interlaminar structure is very pronounced, specifically due to the curved and continuously shaped chitin fibers, which are shown in Fig. 14.

The strengthening effect for the whole specimen that is due to the trabeculae can be evaluated from instances where the trabeculae are or are not present to provide resistance during the peeling process. The whole strengthening effect, λ_m, was calculated using equations (6) and (7):

$$\lambda(\delta) = \frac{F(\delta)}{F_i(\delta)} \geq \frac{F(\delta)}{F_{min0}} \tag{6}$$

$$\lambda_m \geq \frac{\int_{\delta_1}^{\delta_2} F(\delta) d\delta}{\int_{\delta_1}^{\delta_2} F_{min0} d\delta} = \frac{\int_{\delta_1}^{\delta_2} F(\delta) d\delta}{F_{min0} \cdot (\delta_2 - \delta_1)} \tag{7}$$

where $\lambda(\delta)$ is the strengthening effect at a peeling distance, δ. In this paper, the average of 10 specimens, $\overline{F(\delta)}$, $\overline{F_{min0}}$, $\overline{\lambda_m}$ and the standard deviations for $F(\delta)$, $F_{min0}(\delta)$, λ_m were calculated. The value of $\overline{F(\delta)}$ is given by equation (8)

$$\overline{F(\delta)} = \sum_{i=1}^{10} \left[\int_{\delta_1}^{\delta_2} F_i(\delta) d\delta \Big/ (\delta_2 - \delta_1) \right] \Big/ 10 \tag{8}$$

The entire strengthening effect of the interlaminar strength due to the trabeculae in the chitin fiber laminae was approximately 3 times as large as that in the chitin fiber laminae without the trabeculae. Therefore, the trabecular structure of the forewing is not only lightweight but also high-strength.

Object	$\overline{F(\delta)}$ (N)	$\overline{F_{min0}}$ (N)	$\overline{\lambda_m}$
Average	0.148	0.051	3.2
Standard deviation	0.047	0.019	1.1

Table 1. Strength ratio and its standard deviation by trabecula

3.3. Conclusions

A peeling test between the chitin fiber laminae of the *A. dichotoma* forewing was conducted, and the reinforcement mechanism, especially for the 3D interlaminar reinforcement, was investigated with the following results.

1. Many load peaks were observed in the peeling curves with their number being equal to the number of trabeculae found in the chitin fiber laminae. This finding demonstrated that the fracture of each trabecula contributed to a single load peak in the peeling curve, which ultimately contributed to the interlaminar strength. Furthermore, the characteristics of the peeling curve depended on the density and 2D distribution of the trabeculae within the forewing.
2. The chitin fibers were interconnected between the chitin fiber laminae and the trabeculae, and they were curved and continuously shaped where the trabeculae were bonded to the chitin fiber laminae. The fibers were distributed in two dimensions within the plane of each chitin fiber lamina. The chitin fibers broke in tension during the peeling process and exhibited a high resistance to the applied load.
3. The amount of the interlaminar strength that was due to the trabeculae in the chitin fiber laminae was approximately 3 times greater within the whole region and about 30

times greater in the local region compared to that in the chitin fiber laminae without any trabeculae. Based on the above results, this strong natural reinforcement mechanism was understood, and a model for the reinforcement mechanism was proposed.

4. Optimal composite structures in the forewings of beetles

Within the forewings of insects, both the laminated arrangement of the chitin fibers [Gullan & Cranstion, 1994; Leopoldj et al., 1992; Banerjee, 1988; Zelazny & Neville, 1972] and the mechanical characteristics of these equiangular laminating layers of the biomimetic composites [Tanimoto et al., 1998; Ben et al., 1998; Masuda et al., 1995] have been reported on previously. The authors have reported the presence of non-equiangular laminating structures in the forewing of *A. dichotoma* [Chen et al., 2001a]. It is well known that the somatotypes of beetles differ from one another due to their sex, with the larger male bearing a horn. In the present study, based upon previous work on the characteristics of somatotype beetles, the forewing, the tensile fracture force of the forewing and its suitability for biomimetic studies have been examined for male and female beetles. Both optimal structures of the fiber's cross section and the typical reticular cross-linking of two-dimensional reinforced structures within the beetle's forewing are discussed.

4.1. Experimental specimens and methods

Two types of forewing specimens were selected from both male and female *A. dichotoma*, as shown in Fig. 15(a) and Fig. 1(a). The experimental bio-samples included a total of 12 males and 12 females with measurements of each beetle's weight, its forewing dimensions and its physical size.

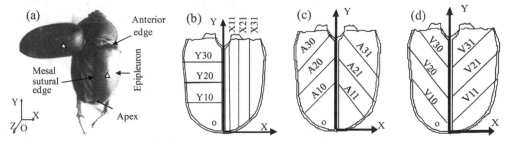

Figure 15. Female beetle (a) and example position in (b) 0° or 90°, (c) 45°, (d) 135° of the beetle forewing.

The different orientations of the tensile test specimens were prepared as shown in Fig. 15. The tensile test specimens measuring 4 mm wide were then sliced and prepared by a special parallel cutting blade [Chen et al., 2001b]. A total of 22 test specimens for both the left and right forewings were prepared. Under the conditions of constant room temperature and humidity (20±2°C, 60%±5%RH), the tensile test was carried out on a Shimazu AUTOGRAPH DSC-10T with a special low load cell and jig at a tensile strain rate of 0.5 mm/min, where the distance between clamps was 5 mm. Additionally, a special clamp head was attached with cloth tape to prevent failure. After the tensile test, the thickness of both the upper and lower lamination layers were measured at four points using the ESEM-2700, and the tensile strength of the specimens were evaluated by taking into account the average thickness of each specimen.

4.2. Experimental results and discussions

4.2.1. Somatotypes

Table 2 shows the measurement results of the somatotypes of *A. dichotoma*, along with the weight and size of the forewing. This also includes the mean values and the standard deviation values (STDEV).

To identify the sex-related differences between the male and female *A. dichotoma*, a T-test [Yoshikawa et al., 1985] was performed on these experimental results, comparing sample group A (male) to sample group B (female). The results of the T-test are shown in Table 2. When the probability value is $P(t_0) \leq \alpha$ (generally $\alpha=0.05$ at normal significance level), there exists a significant difference and a * is labeled in that specific cell of Table 2. It is clear that both the weight and physical size of the male *A. dichotoma* are larger than the weight and size of the female. The male forewing measures 28.7 mm in length and 14.8 mm in width, whereas the female forewing measures only 28.3 mm in length and 13.6 mm in width. The significant differences indicate that the male is larger than the female; however, the forewings of the male and female beetle have a similar weight. In fact, females have a forewing weight of 0.160 g and males have a forewing weight of 0.151 g; surprisingly, the female beetles weigh slightly more than the males, according to this experimental data.

Compared to female *A. dichotoma*, males have a larger somatotype and physical size. However, the male also has a lighter forewing. It is thus supposed that there exist certain physical differences between males and females in *A. dichotoma* forewings in features such as thickness, laminated arrangement and laminated layers. Accordingly, these differences between male and female beetles should be considered when their forewing structures and mechanical properties are investigated.

Beetle	Weight(g)		Length(mm)		Width(mm)	
	Male	Female	Male	Female	Male	Female
AVERAGE	5.58	4.72	46.13	42.82	23.30	21.79
STDEV	0.83	0.97	2.96	2.75	1.14	1.10
$P(t_0)$	0.03	*	0.01	*	0.00	*
Forewing	Male	Female	Male	Female	Male	Female
AVERAGE	0.151	0.160	28.70	28.30	14.80	13.60
STDEV	0.015	0.022	1.26	1.23	0.65	0.56
$P(t_0)$	0.27		0.01	*	0.00	*

Table 2. Measurements of the *A. dichotoma* forewing

4.2.2. Forewing strength and optimal forewing structure according to sex

Table 3 shows the tensile fracture force of the 4-mm-wide male and female specimens, including the thickness of the upper and lower laminations of the forewings and the corresponding stress. The figures above are the mean values obtained for specimens in all orientations as shown in Fig. 15. The T-test results regarding male and female *A. dichotoma* are also presented in Table 3. Although there is no difference in stress between the male and female forewing, there are significant differences in both the thickness and the tensile fracture force. The forewings of the females are thicker and exhibit a higher tensile fracture force than the males.

Beetle	Object	Force	Thickness	Stress
		(KN)	(μm)	(MPa)
Male	Average	28.1	54.0	130.7
	STDEV	5.94	7.22	25.84
Female	Average	35.6	69.7	127.9
	STDEV	2.32	8.04	25.20
T-test	$P(t_0)$	0.00	0.00	0.49
		*	*	

Table 3. Tensile results of the forewing

However, in nature, beetles regularly engage in combat with each other or with insects of other species, either when searching for food or for a mate. For this reason, injured spots can often be found on beetle forewings, which are the result of tearing by the horns of competing beetles. Initially, 20 of the *A. dichotoma* died, most likely due to fighting or other natural causes. Among these beetles, 14 were male and 34 injured spots were identified among the forewings of the 12 wounded beetles, i.e., an average of 2.43 injured spots per male *A. dichotoma* [Chen et al., 2007a]. Meanwhile, only a single injured spot was located on the forewings of the females. Therefore, it is reasonable to assume that the differences in fighting injuries may be the result of evolutionary adaptations.

The male *A. dichotoma* has a heavy, large body with a light forewing (see Table 2). The beetle uses its lightweight wing to support its heavy body effectively in flight and uses its horn to protect its body. Observations have indicated that the male has a large but thin forewing with a lightweight construction. The females, who lack horns, have only the forewing to serve as protecting armor. It is thus reasonable to assume that possession of a thick forewing maximizes the female beetle's chance of survival in nature, increasing its lifespan and mating opportunities and further affording it a better opportunity to pass on its genes to the next generation. As a result of the thickened forewing, the beetle's tensile fracture force is increased, which forms a stronger barrier against external damage (see Table 3).

Furthermore, it is also hypothesized that the forewing of the female *A. dichotoma* is reinforced not only by an increased thickness but also by varying the orientation of its laminated layers and other features. These possible reinforcement mechanisms are areas for future work. The following discussion, however, primarily covers the fiber-reinforcement mechanism and the respective design techniques within both the male and female *A. dichotoma*.

4.2.3. Optimum cross section of the chitin fibers and their volume fraction in the forewing

As discussed above, the chitin fibers play a role in fiber enhancement of the forewing. Here, we discuss the "shape" (cross section) and the "amount" (density as a percentage) of reinforced fiber enhancements within the current bio-composite materials.

To discuss the cross sectional shape of the fibers and the distribution of the fiber density in the forewing, the forewing fiber cross sectional information must first be discussed. As shown in Fig. 16(a, b), the forewing consists of an upper lamination, a lower lamination and a void lamination, in which many trabeculae are distributed throughout. The inside of both the upper and lower laminations, called the endocuticle, is composed of chitin fibers (see Fig. 16a, b). The outsides of the lamination regions, called the exocuticle, are composed mainly of proteins. It was observed that the fibers terminated at the exocuticle of the epipleuron tip, and their cross sectional shapes were different from the fibers located at the endocuticle. Cross sections of the fibers that were observed in the exocuticle and the endocuticle are shown in Fig. 17, below. Also in Fig. 17(a, b), the fibers at the epipleuron tip of the forewing can be seen to have a circular cross section with a sparse distribution. Meanwhile, the fibers located primarily around the void lamination in the forewing's endocuticle are densely distributed with a rectangular cross section as seen in Fig. 17(c, d).

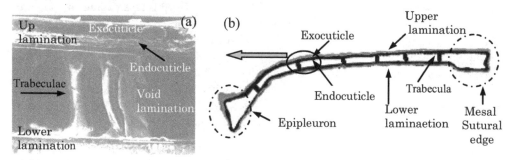

Figure 16. Schematic of section component in the forewing of beetle. (a) micrograph, (b) single model of whole section of forewing.

Fig. 18 illustrates two types of idealized chitin fibers, with either square or circular cross sections. Assuming that both the side of the square and the diameter of the circle are of the same dimension D, then multi-layers of the chitin fibers oriented normally at 90° in a dense and sparse distribution as shown in Fig. 18 (a) and (b). Under the same densely distributed condition, the chitin fiber volume fraction in the square cross section (Fig. 18a) is approximately 27% higher than that of the circular (Fig. 18b). It is thus supposed that the fiber reinforcements with rectangular cross sections result in a maximal fiber volume and the high strength of the forewing.

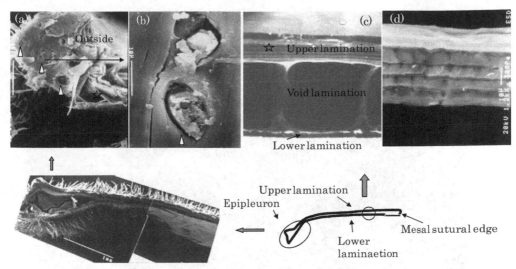

Figure 17. Typical cross sectional shape of the Chitin fibers and their positions in the forewing of *A. dichotoma*, the circular cross section (Δ) (a) and a magnified view (b) within the epipleuron tip and the rectangular cross section (☆) (c) around the void lamination in the endocuticle, as well as its magnified view (d).

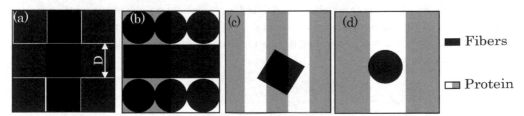

Figure 18. Model of a chitin fiber cross section · Higher density with (a), rectangular and (b) circular shapes and lower density with (c) rectangular and (d) circular shapes.

Next, the sparsely distributed samples of square and circular chitin fibers, as shown in Fig. 18 (c) and (d), have approximately only a 5% fiber volume fraction. The circular samples have less stress concentration than either rectangular or triangular cross sections would, and they also have a more effective interface with the surrounding protein matrix. Furthermore, the circular samples measure 10 μm in diameter and are twice as broad as those of the square cross section in length; its polar moment of inertia is more than ten times higher than that of the rectangular cross sections because it is proportional to the fourth power of the cross section size.

4.2.4. Optimal structures by kind of beetles

For comparison, photographs of *P. inclinatus* (Fig. 19 to Fig. 21) display the sections of the forewing, the trabecular distribution and their structure. Based on these, we confirmed that the fundamental structures of *P. inclinatus*, such as the frame structure (Fig. 19), the chitin fiber layers and their arrangement, are similar to that of *A. dichotoma*. The ratio of the thickness at the edge or the thickness at the central region to the total thickness of both beetles is nearly the same at any position (at the edge or center), but the void thickness in *A. dichotoma* is much larger than that of *P. inclinatus* [Chen et al., 2000].

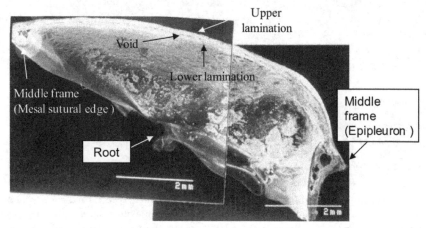

Figure 19. Structure of different sections of the forewing of *P. inclinatus*.

The average density of the trabeculae, calculated as the number of trabeculae per square of mm, is 35 for *P. inclinatus*, and 6 for *A. dichotoma*, which is a difference in trabecular density of approximately six times. The trabeculae can be seen in Fig. 20(a) for *A. dichotoma*, and Fig. 20(b) for *P. inclinatus*.

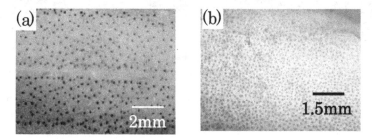

Figure 20. The forewing and trabeculae with 10% KOH treatment of (a) *A. dichotoma*, (b) *P. inclinatus*.

The trabeculae in *A. dichotoma* are also relatively long and of a smaller diameter, whereas in *P. inclinatus*, they are relatively short with a larger diameter (Fig. 21a-c).

Here, the differences in structural details between *A. dichotoma* and *P. inclinatus* are discussed. Although both species of beetles possess a lightweight structure based upon a sandwich plate structure and frame construction with variable cross section, there are still measurable differences that exist between the two. For example, *P. inclinatus* has not only short and large diameter trabeculae but also a trabecular density six times greater than *A. dichotoma*. The forewing of *P. inclinatus* is twice as thick in the upper and lower laminations,

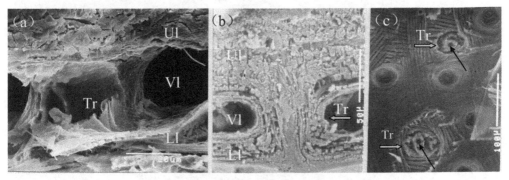

Figure 21. Microstructure of the trabeculae of *P. inclinatus*. (a) whole, (b) in longitudinal section, (c) in upper lamination following treatment with KOH.

and its void height is 40% that of *A. dichotoma* [Chen et al.,2000]. Based on these differences, we consider that *A. dichotoma* must have a weaker forewing structure, while *P. inclinatus* possesses the higher strength structure. The forewing structure of *P. inclinatus* is optimized mainly by its lightweight structure, which is achieved by varying the quantity parameters, such as trabecular diameter, density, and thickness of lamination.

Why do these two beetles have such obvious structural differences? We have considered that these differences may be related to the lifespan of each beetle species. *P. inclinatus* has a longer lifespan and is able to survive through the winter, but *A. dichotoma* has a lifespan of only a few weeks and cannot survive the winter. In other words, *P. inclinatus* would be expected to have more durable physical characteristics than *A. dichotoma* because the latter has an extremely short lifespan. From a product viewpoint, the structural design of *P. inclinatus* is meant for durability and longevity, while *A. dichotoma* is of a more economical design with shorter lifetime. It is also noteworthy that the middle section of the lower lamination of *A. dichotoma* is extremely thin, possessing a thickness of only 1/3 that of *P. inclinatus*. Because this part of the beetle's body is unexposed and thus rarely attacked by enemies in nature, the evolution of an extremely lightweight forewing in *A. dichotoma* resulted [Chen et al., 2007b].

4.3. Conclusions

In this section, the optimal composite structures in the beetle's forewing were discussed based on the beetle's sex, the tensile fracture force of the forewing, the chitin fiber shape and the fiber content. The results are as follows:

1. In comparison to the females, it is clear that the males have lighter and thinner forewings with a lower tensile fracture force; this is because male beetles have a large horn to protect them, while the females do not. The females, with their larger bodies to produce offspring · have only their thick forewings to serve as protecting armor.
2. By investigating the characteristics of the reinforced composite structures within the beetle forewing, it was found that densely distributed chitin fibers were located around

the void lamination in the endocuticle of the forewing and that they have a rectangular cross section, possessing a maximum fiber volume fraction for reinforcement. Meanwhile, the sparsely distributed cross sections within the exocuticle at the epipleuron tip of the forewing are of a circular cross section and have a strong interfacial reinforcement within a protein matrix.

3. Both beetle species forewings are of the sandwich plate structure with a central void layer, in which there exists many distributed trabeculae. Thus, the beetle forewings are a lightweight composite frame structural design. The structure of the *A. dichotoma* forewing is disposable and shows an economical design, while the structure of *P. inclinatus* shows a strong and durable design.

5. Applications of the beetle forewing structure for developing composite materials

Honeycomb structures are typical lightweight and high-strength composites. There are numerous examples of related studies and applications: from nanomaterials [Kim, 2006; Hideki & Kenji, 1995] to massive structures, such as those used in F4 fighters, Boeing-767 and Airbus-380 airplanes, and Boeing-360 helicopters [Sato & Kino, 2004; Shafizadeh et al., 1999; Llorente, 1989]. Presently manufactured commercial honeycomb sandwich plates are produced by adhesively joining the plate parts and core parts, which are made separately by different processes [Leng, 2009; http://www.corecomposites.com; Chen et al., 2008; http://www.nida-core.com/english/contact.htm]. The sandwich plates made by the above processes can be readily separated at the joint of the side plates to the core, and this separation potential is a factor that limits both the strength and the side sealing effectiveness. Furthermore, the use of the adhesive glue is not only environmentally harmful but also expensive [Han et al., 2002; Gu et al., 2010]. To overcome the weaknesses associated with traditional manufacturing method, we have recently developed an integrated molding process [Chen et al., 2012] and the bionic composite sample of the first integrated honeycomb plate is shown in Fig. 22

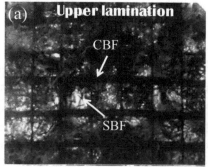 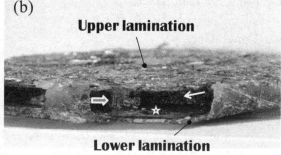

Figure 22. An example of the bionic composite. (a) Front view, (b) side view. CBF denotes continuous basalt fiber; SBF denotes short basalt fiber; the thick and thin arrows indicate the trabeculae and the honeycomb walls, respectively, and the star indicates a processing hole.

Fig. 23 shows the flow chart of the integrated manufacturing process used to produce the honeycomb plate [Chen et al., 2012]. To improve the association between the BF (Basalt Fiber) and the epoxy resin, the surfaces of the short BFs and CBF (Continuous Basalt Fiber) geogrid were treated with a silane-coupling agent. At a mass ratio of 30%, the BFs were placed in an ethanol solution with 0.75% KH550 (γ-aminopropyltriethoxysilane); the BFs were removed from the ethanol solution after 30 min of soaking and were allowed to dry naturally. Finally, the BFs were heat treated for 1 hour at 120 °C in an oven. At the same time, the mold tools were prepared for the integrated molding, the CBF geogrid was placed in the upper and lower layer, and then the short BFs and epoxy resin were evenly mixed. Table 4 lists the materials used for the mixture and their mass ratios [Chen et al., 2012].

5.1. Experiment

Fig. 23 shows the flow chart of the integrated manufacturing process used to produce the honeycomb plate [Chen et al., 2012]. To improve the association between the BF (Basalt Fiber) and the epoxy resin, the surfaces of the short BFs and CBF (Continuous Basalt Fiber) geogrid were treated with a silane-coupling agent. At a mass ratio of 30%, the BFs were placed in an ethanol solution with 0.75% KH550 (γ-aminopropyltriethoxysilane); the BFs were removed from the ethanol solution after 30 min of soaking and were allowed to dry naturally. Finally, the BFs were heat treated for 1 hour at 120 °C in an oven. At the same time, the mold tools were prepared for the integrated molding, the CBF geogrid was placed in the upper and lower layer, and then the short BFs and epoxy resin were evenly mixed. Table 4 lists the materials used for the mixture and their mass ratios [Chen et al., 2012].

After pouring the well-mixed fluid of fibers and resin into the mold tools, the mixture was vacuum-dried for 30 minutes at room temperature in a vacuum oven. A portion of aluminum foil and a heavy object were set on the assembly of mold tools after the tools were removed from the vacuum oven. The assembly was then cured for 10 hours at 35 °C in an incubator. The final molded sample was obtained by removing the paraffin support at 75 °C in the incubator.

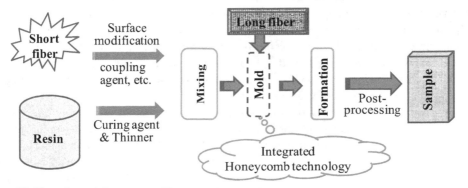

Figure 23. Flow chart of the integrated honeycomb technology.

Material	Epoxy resin	Curing agent	Thinner	Short basalt fiber
m/g	80	20	10	10
w/%	66.7	16.7	8.3	8.3

Table 4. Main materials and their mass ratios

5.2. Results and discussion

In this section, we begin with the examination of the natural edge sealing structure used in the beetle forewing, then we investigate the mechanisms of structure formation and the structure functions, and last, we propose an edge sealing technique for integrated honeycomb processing technology.

5.2.1. Edge sealing of the forewing integrated honeycomb plates and its biomimetic application

As mentioned above, the macroscopic structure of the beetle forewing features a sandwich plate construction with extensive internal meshwork (denoted by stars in Fig. 2 a, b) and hundreds of trabeculae (denoted by thick arrows in Fig. 2 a-e). The cross section of the forewing edge frame changes by location (denoted by EFS in Fig. 2a, e). With the exception of the concavo-convex junction between the left (denoted by round arrows, Fig. 2e) and the right forewings, the edge frame is an ingeniously integrated sealing structure without any seams (Fig. 2e). Before the bionic application of the forewing of A. dichotoma could be carried out, a search for the formation mechanisms and an analysis of the structural function were conducted, as discussed in the next section.

5.2.2. The formation mechanisms of the forewing integrated trabecular honeycomb structure

To understand the natural mechanisms behind the formation of the forewing integrated honeycomb structure, we first investigated the hexagonally shaped honeycomb structure inside the beetle forewings (Fig. 2a, b); specifically, we were interested in the edge sealing structure without any seams. According to Thompson's cell partitioning theory (Fig. 24) [Thompson, 1945], when a cell membrane intersects with two other cells on a plane, angles α, β, and γ are determined by the reciprocity of the three tensions t, T, and T' at the intersection O. For example, three scenarios are shown in Fig. 24 (a)-(c) as follows: (a) when T'=T>>t, i.e., when comparing T' and T, t can be ignored, and in this case, α and β are right angles; (b) when T'=T>t, i.e., T' and T are quite large compared with t, and thus, α and β are obtuse angles; (c) α, β, and γ are all 120° when T'=T=t. The mesh structures inside the forewing of the A. dichotoma beetle are extensively distributed, and the forewing exhibits a curved structure with thousands of trabeculae and air sacs possessing dissimilar internal stresses. Most of the cells of the forewing have essentially the same function as the branched veins of the dragonfly wing [Thompson, 1945], and the forewing air sacs are distributed in a specific manner to form a 120° angle between the two adjacent membranes, i.e., they are similar to a honeycomb structure (Fig. 2b).

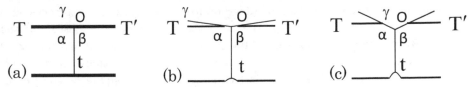

Figure 24. The theoretical cell shape and the relationship between the three tensile forces T', T and t: (a) T'=T≫t, (b) T'=T>t, (c) T'=T=t [Thompson, 1945].

In addition, to better understand why the forewing is constructed with the trabeculae and edge frame of the honeycomb structure, the functions of these structural features, the trabeculae and the edge frame, have been investigated previously [Chen et al., 2000; Chen et al., 2001a; Chen et al., 2007b; Ni et al., 2001]. These investigations have reported the following: the trabeculae and the edge frame provide bending stiffness and resistance to applied stress for the entire forewing structure [Chen et al., 2000; Chen et al., 2007b]. These structures also effectively increase the inter-laminar strength between the lamination layers of the composite structure [Chen et al., 2001a; Ni et al., 2001]. Furthermore, the structures impart the forewing with sufficient mechanical properties to enable the beetle with flight capabilities. These beneficial properties are achieved by the specific angles held between the different fiber layers in the upper lamination of the forewing [Chen et al., 2001a] and the natural design of the transitional region of the edge frame [Chen et al., 2000]. In this manner, the entire forewing of the *A. dichotoma* beetle forms the completely integrated trabecular honeycomb structure.

5.2.3. Integrated honeycomb plates with edge sealing and the bionic imitation technique

In the previous section, a complete schematic of a beetle forewing was included to demonstrate the structure of the integrated trabecular honeycomb plate with edge sealing (Fig. 25); in the present work, we did not consider the special functions of the edge frame,

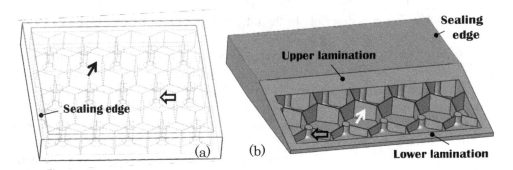

Figure 25. The edge sealing schematic developed for manufacturing the integrated honeycomb, which was inspired by the beetle forewing. (a) Entire integrated honeycomb plate, (b) oblique cross section of (a). The thick arrows indicate trabeculae; the thin arrows indicate honeycomb wall segments.

and thus, the hollow part of the edge frame was not incorporated. This structure helped us to develop not only a new manufacturing method for producing honeycomb plates but also a new edge sealing technology. Fig. 26a shows a set of mold tools used for sealing operations. By adding extra space around the edge of the female mold [Chen et al., 2012], a sealing edge can be generated. Fig. 26b shows a cross sectional schematic of the sealing edge of the integrated honeycomb plate. This technology does not require bonding or other mechanical connections made by nail or inlay and is therefore preferred over the traditional honeycomb plates because of the simplified manufacturing process, the perfect sealing (as seams are not required) and the enhanced integration.

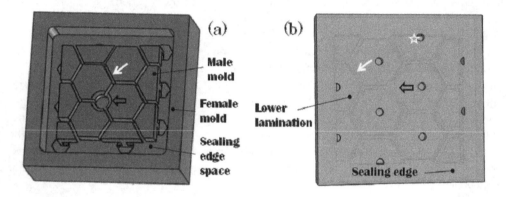

Figure 26. Schematic of the edge sealing approach for use in the integrated honeycomb technology. (a) Mold tools for bionic sealing, (b) image of the integrated honeycomb plate with the sealing edge. The thick arrows indicate trabeculae or the space used to form it; the thin arrows indicate honeycomb wall segments or the space used to form it, and the star indicates a processing hole.

5.2.4. Processing and optimization of the integrated honeycomb technology

Optimization of the processing hole In the integrated honeycomb plate, as shown in Fig. 26, there is a small processing hole in each core. However, the processing holes can have an adverse effect on the strength of the composite and should ultimately be minimized or eliminated. Certain approaches, such as decreasing the diameter of the holes or reducing their number by using integrated male molds (Fig. 27a), may prove effective toward reducing the impact of the processing holes on the composite strength. Increasing the interconnections between the basic male molds to integrate them into the structure can reduce the number of the processing holes needed; however, this requires special molds, as shown in Fig. 27b, c. To accomplish this goal, a thin-walled hexagonal box for each honeycomb unit can be formed and then integrated [Chen et al., 2012]. Integrating the boxes and then fusing them into the plate can help to eliminate the number of processing holes required.

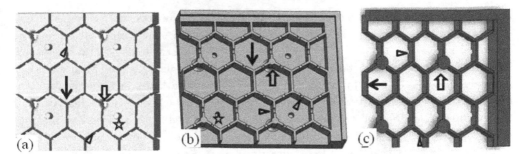

Figure 27. Schematic of the integrated male mold and its mold tools (only a quarter of the symmetric structure is shown in this figure). (a) Bottom view of the integrated male mold, (b) lower half of a mold tool assembly, (c) upper half of a mold tool assembly. The thick arrows show the trabeculae or the space to form them. The thin arrows indicate the honeycomb wall or the space to form them. The star indicates either a positioning hole or a positioning pin. The triangles indicate the processing columns or the space to form them.

Optimization of materials and manufacturing processes As discussed above, the long molding time required for the first method is a drawback. Even so, the molding time can be optimized by selecting appropriate resins or even by using the presently presented process of wax mold tools. For example, two laminating resins, TG-1001™ and TG-1009™, can be cured in a short period of time (such as 5 min) and under a relatively low temperature (33–85 °C) [http://www.thermalguard-technology.com]; resin transfer molding and vacuum-assisted molding are both performed in this temperature range. Such resins, which are noncombustible, exhibit high strength, are refractory, and meet our complete list of requirements. They also have potential for use in furniture manufacturing, civil engineering, transportation devices, and other applications. The disadvantages of these resins, such as the low content and uneven distribution of fibers, can still be optimized by depositing more fiber-woven roving, by prefilling a certain amount of short fibers into the resin, or by applying vacuum-assisted molding processes.

5.2.5. Mechanical properties of the integrated trabecular honeycomb structure

To gain knowledge of the mechanical properties of the integrated trabecular honeycomb structure, we focused on revealing whether fiber interconnections exist between the upper/lower laminations and the trabeculae; these interconnections are considered to be the main difference between the trabecular honeycomb plate and the traditional honeycomb plate. Several studies have previously investigated the mechanical properties of the trabecular honeycomb structure of the beetle forewing as well as bionic materials based on the beetle forewing. First, experiments and analyses using the finite-element method (FEM) were conducted on living forewing specimens and their peeling models; the results indicated that the forewing possesses a high inter-laminar strength value [Ni et al., 2001; Gu et al., 2010]. Second, a bionic composite specimen made from a sandwich structure with fiber-reinforced trabeculae was constructed; the experimental results demonstrated that the energy release rate and the maximum shear stress of the fiber-reinforced material were approximately four times greater than those of a traditional plate without fiber-reinforced trabeculae [Okazaki et al., 2005]. Third, the energy-absorbing capabilities of the original and improved models of the beetle forewing were analyzed using nonlinear FEM, and it is clear from the models that the forewing structure is both crashworthy and energy-absorbing, particularly in the improved models [Guo & Wang, 2011]. Furthermore, the trabecular honeycomb plates that were produced by our bionic method not only achieved integrated preparation between the upper and lower lamination and the core lamination but also produced a natural seamless edge using the new sealing technology. Additionally, the reinforcing fibers were distributed between the core layer (trabeculae, honeycomb walls) and the upper and lower laminations (including the edge part); that is, the bionic materials of the real integration trabecular honeycomb plate are similar to the biological structure of the beetle forewing. Therefore, the bionic plate should exhibit excellent mechanical properties, as suggested by the aforementioned physical experiments and FEM analysis.

5.3. Conclusions

The processes for manufacturing integrated honeycomb plates have been outlined, and the problems associated with the integrated honeycomb technology have been discussed and resolved:

1. Each forewing of a beetle has a natural edge sealing design with a completely integrated structure consisting of honeycomb cores, including trabeculae and an edge frame. The formation mechanism of the integrated trabecular honeycomb structure and its mechanical properties were discussed. An edge sealing technique inspired by the forewing structure was proposed. This technique has numerous advantages, such as processing simplicity and complete seamless integrated.
2. For the first time, detailed manufacturing processes have been developed and presented toward achieving integrated honeycomb plates that include edge sealing. Optimization methods have been developed, such as integrating the male molds that can eliminate the influence of the processing holes and choosing a fast-curing and high-temperature resin that can shorten the molding time.

Author details

Jinxiang Chen* and Juan Xie
International Institute for Urban Systems Engineering & School of Civil Engineering, Southeast University, China

Qing-Qing Ni
Dept. of Functional Machinery & Mechanics, Shinshu University, Japan

---

* Corresponding Author

Acknowledgement

This work was supported by the Natural Science Foundation of China (Grant No. 51173026), the Jiangsu NSF (No.BK2010015) and the National Key Technologies R&D Program of China (2011BAB03B10).

6. References

Banerjee, S. (1988) Organisation of Wing Cuticle in Locusta Migratoria Linnaeus, Tropidacris Cristata Linnaeus and Romalea Microptera Beauvais (Orthoptera: Acrididae). *International Journal of Insect Morphology and Embryology*, Vol. 17, pp 313-326.

Ben, G.; Nishi, Y. & Asano, S. (1998) A Discussion of Mechanical Propreties for an Optimum Stackling Sequence in Beetles. *Nippon Kikai Gakkai Zairyo Rikigaku Bumon Koenkai Koen Ronbunshu*, Vol. 1998-B, pp 349-350.

Chen, J.; Dai, G.; Xu, Y. & Iwamoto, M. (2007a) Optimal Composite Structures in the Forewings of Beetles. *Journal of Composite Structures*, Vol. 81, pp 432-437.

Chen, J.; Gu, C.; Guo, S.; Wan, C.; Wang, X.; Xie, J. & Hu, X. (2012) Integrated Honeycomb Technology Motivated by the Structure of Beetle Forewings. *Journal of Bionic Engineering*. doi:10.1016/j.msec.2012.04.067.

Chen, J.; Iwamoto, M.; Ni, Q.Q.; Kurashiki, K. & Saito, K. (2000) Cross Sectional Structure and its Optimality of the Forewing of Beetles. *Journal of the Society of Materials Science*, Vol. 49, No. 4, pp 407-412.

Chen, J.; Iwamoto, M.; Ni, Q.Q.; Kurashiki, K. & Saito, K. (2001a) Laminated Structure and its Mechanical Properties of the Forewing of Beetle. *Journal of the Society of Materials Science*, Vol. 50, pp 455-460.

Chen, J.; Ni, Q.Q.; Endo, Y. & Iwamoto, M. (2001b) Fine Structure of the Trabeculae in the Forewing of Allomyrina Dichotoma (Linne) and Prosopocoilus Inclinatus, (Motschulskey) (Coleoptera: Scarabaeidae). *Insect Science*, Vol. 8, pp 115-123.

Chen, J.; Ni, Q.Q; Endo, Y. & Iwamoto, M. (2002) Distribution of the Trabeculae in the Forewing of Horned Beetle, Allomyrina Dichotoma (Linne) (Coleoptera: Scarabaeidae). *Insect Science*, Vol. 9, pp 55-61.

Chen, J.; Ni, Q.Q.; Li, Q. & Xu, Y. (2005) Biomimetic Lightweight Composite Structure with Honeycomb-trabecula. *Acta Materiae Compositae Sinica*, Vol. 22, pp 103-108.

Chen, J.; Ni, Q.Q.; Xu, Y. & Iwamoto, M. (2007b) Lightweight Composite Structures in the Forewings of Beetles. *Journal of Composite Structures*, Vol. 79, pp 331-337.

Chen, J.; Xie, J.; Zhu, H.; Guan, S.; Wu, G.; Noori, M.N. & Guo, S. (2012) Integrated Honeycomb Structure of a Beetle Forewing and its Imitation. *Materials Science & Engineering C*, Vol. 32, pp 613-618.

Chen, X.; Sun, Y. & Gong, X. (2008) Honeycomb Textile Composites, and Experimental Analysis of 3D Honeycomb Textile Composites Part I: Design and Manufacture. *Textile Research Journal*, Vol. 78, No. 9, pp 771-781.

Gu, R.; Guo, K.; Qi, C.; Zhang, J. & Liu, Z. (2010) Properties and Bonding Mechanism of Konjak Powder-chitosan-PVA Blending Adhesive. *Transactions of the CSAE*, Vol. 26, No. 5, pp 373-378.

Gullan, P.J. & Cranstion, P. (1994) The Insects: An Outline of Entomology. *Chapman & Hall*, pp 22–55.

Guo, T.; Wang, Y.F.; (2011) Energy Absorbing Structures Imitating Trabecular of Beetle Cuticles. *Engineering Mechanics*, Vol. 28, No. 2, pp 246-251.

Han, T.S.; Ural, A.; Chen, C.S.; Zehnder, A.T.; Ingraffea, A.R.; Billington, S.L. (2002) Delamination Buckling and Propagation Analysis of Honeycomb Panels Using a Cohesive Element Approach. *International Journal of Fracture*, Vol. 115, pp 101-123.

He, M. & Cox, B.N. (1998) Crack Fridging by Through-Thickness Reinforcement in Delaminating Curved Structures, *Composites Part A*, Vol. 29A, pp 377-393.

Hepburn, H.R. (1976) The Insect Integument. *Elsevier scientific Publication Company,New York*, pp 1-100.

Hideki, M. & Kenji, F. (1995) Ordered Metal Nanohole Arrays Made by a Two-Step Replication of Honeycomb Structures of Anodic Alumina. *Science*, Vol. 268, No. 5216, pp 1466-1468.

Hosur, M.V.; Abdullah, M.; Jeelani, S. (2005) Manufacturing and low-velocity impact characterization of foam filled 3-D integrated core sandwich composites with hybrid face sheets. *Composite Structures*, Vol. 69, No. 2, pp 167-181.

http://www.corecomposites.com

http://www.thermalguard-technology.com.

http://www.nida-core.com/english/contact.htm

Ishii, S. (1982) Physiology of Insects. *Baifukan press*, pp 40-57.

Iwamoto, M.; Chen, J.; Ni, Q.Q.; Kurashiki, K. & Saito, K. (1999b) Structure Optimality with Biomimetics of the Fore-wing of Beetles. *The 3rd China-Japan Joint Conference on Composites, HangZhou*, pp 15-20.

Iwamoto, M.; Ni, Q.Q.; Fujwara, T. & Kurashiki, K. (1999a) Interlaminar Fracture Mechanism in Unidirectional CFRP Composites-Part I: Interlaminar Toughness and AE Characteristics. *Engineering Fracture Mechanics*, Vol. 64, pp 721-745 .

Jung, W.Y. & Aref, A.J. (2005) Analytical and Numerical Studies of Polymer Matrix Composite Sandwich Infill Panels. *Composite Structures*, Vol. 68, pp 359-370.

Kim, H.; Kim, J.; Yang, H.; Suh, J.; Kim, T.; Han, B.; Kim, S.; Kim, D.S.; Pikhitsa, P.V. & Choi, M.;. (2006) Parallel Patterning of Nanoparticles via Electrodynamic Focusing of Charged Aerosols. *Nature Nanotechnology*, Vol. 1, pp 117-121.

Leopoldj, R.A.; Newman, S.M. & Helgeson, G. (1992) A Comparison of Cuticle Deposition during the Pre-and Post-eclosion Stages of the Adult Weevil, Anthonomus Grandis Boheman. *International Journal of Insect Morphology and Embryoogyl*, Vol. 21, pp 37-62.

Llorente, S. (1989) Honeycomb Sandwich Primary Structure Applications on the Boeing Model 360 Helicopter. *Society for the Advancement of Material and Process Engineering*, Vol. 34, No. 5, pp 824-838.

Mania, R. (2005) Buckling Analysis of Trapezoidal Composite Sandwich Plate Subjected to In-plane Compression. *Composite Structures*, Vol. 69, pp 482-490.

Masuda, Y.; Okada, A.; Tabata, H.; Yoneda, K. & Yamamoto, Y. (1995) Structure and mechanical properties of the elytra of pachyrhynchid weevils. *Lecture Article of L,ecture Meeting of Japan Mechanical Society Department of Materials Mechanics*, pp 121-122.

Matsunaga, H. (2002) Assessment of a Global Higher-order Deformation Theory for Laminated Composite and Sandwich Plates. *Composite Structures*, Vol. 56, pp 279-291.

Nguyen, V.T.; Caron, J.F. & Sab, K. (2005) A Model for Thick Laminates and Sandwich Plates. *Composites Science and Technology*, Vol. 65, pp 475-489.

Ni, Q.Q.; Chen, J.; Iwamoto, M.; Kurashiki, K. & Saito, K. (2001) Interlaminar Reinforcement Mechanism in a Beetle Fore-Wing. *JSME International Journal Series C*, Vol. 44, No. 4, pp 1111-1116.

Okazaki, J.; Ni, Q.Q; Iwamoto, M. & Kurashiki, K. (2005) Research on Design of a Composite Material Imitated from the Fore-wing of a Beetle. *Third International Workshop on Green Composites*, pp 91-94.

Sato, T. & Kino, Z. (2004) Exploration for application business of honeycomb cores and composite materials. *Function and Materials*, Vol. 24, pp 64-73.

Sekine, H. & Kamiya, S. (1987) Analysis of a Shear Fracture in Unidirectional Fiber-Reinforced Composites II (For the Case That the Displacement on Boundaries is Constrained in the Direction Normal to the Shear Fracture). *Japan Society Mechanical Engineers*, Vol. 53, No. 489A, pp 930-934.

Shafizadeh, J.E; Seferis, J.C.; Chesmar, E.F. & Geyer, R. (1999) Evaluation of the in-service performance behavior of honeycomb composite sandwich structures. *Journal of Materials Engineering and Performance*, Vol. 8, No. 6, pp 661-668.

Tanimoto, T.; Chen, Q.H. & Taguchi, M. (1998) Damage Tolerance of biomimetic CFRP Laminate. *Abstacts of 51th Annual Meeting of Textile Mechanical Society*, pp 78-79.

Thompson, D. W. (1945) On Growth and Form. *Cambridge University press*, pp 1-131.

Turss, R.W.; Hine, P.J. & Duckett, R.A. (1997) Interlaminar and Intralaminar Fracture Toughness of UniaxialContinuous and Discontinuous Carbon Fiber/Epocy Composites. *Composites Part A: Applied Science and Manufacturing*, Vol. 28, pp 627-636.

Leng, L. (2009) An Edge-sealed Table Top Board with Paper Honeycomb Core. Patent (Uk) No. 0722071.8.

Wainwright, S.A.; Biggs, W.D.; Currey, J.D. & Gosline, J.M. (1976) Mechanical Design in Organisms. *Princeton University Press*, pp 158-170.

Yoshikawa, H. (1985) The Statistical Analysis Procedure by the Portable Calculator. *Japan Union Science Engineering Press publisher*, pp 28-32.

Zelazny, B. & Neville, A.C. (1972) Quantitative Studies on Fibril Orientation in Beetle Endocuticle. *Insect Physiology*, Vol. 18, pp 2095-2121.

Chapter 6

Comparative Review Study on Elastic Properties Modeling for Unidirectional Composite Materials

Rafic Younes, Ali Hallal, Farouk Fardoun and Fadi Hajj Chehade

Additional information is available at the end of the chapter

1. Introduction

Due to the outstanding properties of 2D and 3D textile composites, the use of 3D fiber reinforced in high-tech industrial domains (spatial, aeronautic, automotive, naval, etc…) has been expanded in recent years. Thus, the evaluation of their elastic properties is crucial for the use of such types of composites in advanced industries. The analytical or numerical modeling of textile composites in order to evaluate their elastic properties depend on the prediction of the elastic properties of unidirectional composite materials with long fibers composites "UD". UD composites represent the basic element in modeling all laminates or 2D or 3D fabrics. They are considered as transversely isotropic materials composed of two phases: the reinforcement phase and the matrix phase. Isotropic fibers (e.g. glass fibers) or anisotropic fibers (e.g. carbon fibers) represent the reinforcement phase while, in general, isotropic materials (e.g. epoxy, ceramics, etc…) represent the matrix phase (Figure 1).

The effective stiffness and compliance matrices of a transversely isotropic material are defined in the elastic regime by five independent engineering constants: longitudinal and transversal Young's moduli E_{11} and E_{22}, longitudinal and transversal shear moduli G_{12} and G_{23}, and major Poisson's ratio ν_{12} (Noting that direction 1 is along the fiber). The minor Poisson's ratio ν_{23} is related to E_{22} and G_{12}. The effective elastic properties are evaluated in terms of mechanical properties of fibers and matrix (Young's and shear moduli, Poisson's ratios and the fiber volume fraction V^f). The compliance matrix [S] of a transversely isotropic material is given as follow:

$$[S] = \begin{bmatrix} 1/E_{11} & -\nu_{12}/E_{11} & -\nu_{12}/E_{11} & 0 & 0 & 0 \\ -\nu_{12}/E_{11} & 1/E_{22} & -\nu_{23}/E_{22} & 0 & 0 & 0 \\ -\nu_{12}/E_{11} & -\nu_{23}/E_{22} & 1/E_{22} & 0 & 0 & 0 \\ 0 & 0 & 0 & 1/G_{23} & 0 & 0 \\ 0 & 0 & 0 & 0 & 1/G_{12} & 0 \\ 0 & 0 & 0 & 0 & 0 & 1/G_{12} \end{bmatrix}$$

The stiffness matrix [C] is the invers of the compliance matrix [S].

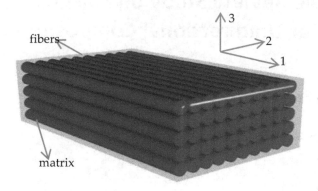

Figure 1. Unidirectional Composite.

In this chapter, a review of most known available analytical micromechanical models is presented in the second section of this chapter. Investigated models belonged to different categories: phenomenological models, semi-empirical models, elasticity approach models and homogenization models. In addition, the evaluation of elastic properties of UD composites using numerical FE method is investigated. Boundary, symmetric and periodic conditions, with different unit cells (square, hexagonal and diamond arrays), are discussed. In the third section, a comparison of the results obtained by the investigated analytical and numerical models is compared to available experimental data for different kinds of UD composites.

2. Review

The prediction of the mechanical properties of UD composites has been the main objective of many researches. Various micromechanical models have been proposed to evaluate the elastic properties of UD composites. These models could be divided into four categories: phenomenological models, elasticity approach models, semi-empirical models and homogenization models.

2.1. Phenomenological models

2.1.1. Rule of Mixture "ROM"

The well-known models that have been proposed and used to evaluate the properties of UD composites are the Voigt [1] and Reuss [2] models. The Voigt model is also known as the rule of mixture model or the iso-strain model, while the Reuss model is also known as the invers rule of mixture model or the iso-stress model.

Elastic properties are extracted from the two models where they are given under the rule of mixture (ROM) and the invers rule of mixture models (IROM).

$$E_{11} = V^f . E_{11}^f + V^m . E^m \quad \text{(from Voigt model)}$$

$$v_{12} = V_f \, v_{11}^f + V_m \, v^m \quad \text{(from Voigt model)}$$

$$E_{22} = \frac{E_{22}^f . E^m}{E^m . V^f + E_{22}^f . V^m} \quad \text{(from Reuss model)}$$

$$G_{12} = \frac{G_{12}^f . G^m}{G^m . V^f + G_{12}^f . V^m} \quad \text{(from Reuss model)}$$

2.2. Semi-empirical models

Semi empirical models have emerged to correct the ROM model where correcting factors are introduced. Under this category, it's noticed three important models: the modified rule of mixture, the Halpin-Tsai model [3] and Chamis model [4].

2.2.1. Modified Rule of Mixture (MROM)

While the investigations show that the obtained results by the ROM model for E_{11} and v_{12} are in good agreement with experimental and finite element data, the results for E_{22} and G_{12} do not agree well with experimental and finite element data. Corrections have been made for E_{22} and G_{12}.

$$\frac{1}{E_{22}} = \frac{\eta^f . V^f}{E_{22}^f} + \frac{\eta^m . V^m}{E^m}$$

Where factors η^f, η^m are calculated as:

$$\eta^f = \frac{E_{11}^f . V^f + \left[\left(1 - v_{12}^f . v_{21}^f\right). E^m + v^m . v_{21}^f . E_{11}^f\right]. V^m}{E_{11}^f . V^f + E^m . V^m}$$

$$\eta^m = \frac{\left[\left(1 - v^{m^2}\right). E_{11}^f - \left(1 - v^m . v_{12}^f\right). E^m\right]. V^f + E^m . V^m}{E_{11}^f . V^f + E^m . V^m}$$

$$\frac{1}{G_{12}} = \frac{\frac{V^f}{G_{12}^f} + \frac{\eta' . V^m}{G^m}}{V^f + \eta' . V^m}$$

With $0 < \eta' < 1$, (it is preferred to take $\eta' = 0.6$)

2.2.2. Halpin–Tsai model [3]

The Halpin-Tsai model also emerged as a semi-empirical model that tends to correct the transversal Young's modulus and longitudinal shear modulus. While for E_{11} and v_{12}, the rule of mixture is used.

$$E_{22} = E^m \cdot \left(\frac{1+\zeta\eta V_f}{1-\eta V_f}\right) \; ; G_{12} = G^m \cdot \left(\frac{1+\zeta\eta V_f}{1-\eta V_f}\right)$$

$$\text{with } \eta = \left(\frac{M_f/M_m - 1}{M_f/M_m + \zeta}\right)$$

with $\zeta = 1$ and 2, and M = E or G for E_{22} and G_{12} respectively.

2.2.3. Chamis model [4]

The Chamis micromechanical model is the most used and trusted model which give a formulation for all five independent elastic properties. It's noticed in this model that E_{11} and v_{12} are also predicted in the same maner of the ROM model, while for other moduli, V^f is replaced by its square root.

$$E_{11} = V^f \, E_{11}^f + V^m \, E^m$$

$$E_{22} = \frac{E^m}{1 - \sqrt{V^f}\left(1 - E^m/E_{22}^f\right)}$$

$$v_{12} = V^f \, v_{12}^f + V^m \, v^m$$

$$G_{12} = \frac{G^m}{1 - \sqrt{V^f}\left(1 - G^m/G_{12}^f\right)}$$

$$G_{23} = \frac{G^m}{1 - \sqrt{V^f}\left(1 - G^m/G_{23}^f\right)}$$

2.3. Elasticity approach models

Under this category, Hashin and Rosen [5] initially proposed a composite cylinder assemblage model (CCA) to evaluate the elastic properties of UD composites. Moreover, Christensen proposed a generalized self-consistent model [6] in order to better evaluate the transversal shear modulus G_{23}.

$$E_{11} = V^f \, E_{11}^f + V^m \, E^m + \frac{4.V^f.V^m.(v_{12}^f - v^m)^2}{\frac{V^f}{K^m} + \frac{1}{G^m} + \frac{V^m}{K^f}} \quad \text{(Hashin and Rosen [5])}$$

$$v_{12} = V^f.v_{12}^f + V^m v^m + \frac{V^f V^m \left(v_{12}^f - v^m\right)\left(\frac{1}{K^m} - \frac{1}{K^f}\right)}{\frac{V^f}{K^m} + \frac{1}{G^m} + \frac{V^m}{K^f}} \quad \text{(Hashin and Rosen [5])}$$

$$G_{12} = G^m . \frac{G^f.(1+V^f) + G^m.V^m}{G^f.V^m + G_m(1+V^f)} \text{(Hashin and Rosen [5])}$$

G_{23} is the solution of the following equation: (Christensen [6])

$$A\left(\frac{G_{23}}{G_m}\right)^2 + 2B\left(\frac{G_{23}}{G_m}\right) + C = 0$$

With:

$$A = 3Vf.\left(1 - Vf\right)^2.\left(\frac{G_{23}^f}{V^m} - 1\right)\left(\frac{G_{23}^f}{G^m} + \eta_f\right)$$

$$+ \left[\frac{G_{23}^f}{G^m}\eta_m + \eta_f\eta_m - \left(\frac{G_{23}^f}{G^m}\eta_m - \eta_f\right)V^{f^3}\right].\left[V_f\eta_m\left(\frac{G_{23}^f}{G^m} - 1\right) - \left(\frac{G_{23}^f}{G^m}\eta_m + 1\right)\right]$$

$$B = -3V^f V^{m^2}\left(\frac{G_{23}^f}{G^m} - 1\right)\left(\frac{G_{23}^f}{G^m} + \eta_f\right) + \frac{V^f}{2}(\eta_m + 1)\left(\frac{G_{23}^f}{G^m} - 1\right)\left[\frac{G_{23}^f}{G^m} + \eta_f + \left(\frac{G_{23}^f}{G^m}\eta_m - \eta_f\right)V^{f^3}\right]$$

$$+ \left\{\frac{1}{2}\left[\frac{G_{23}^f}{G^m}\eta_m - \left(\frac{G_{23}^f}{G^m} - 1\right)V^f + 1\right].\left[(\eta_f - 1)\left(\frac{G_{23}^f}{G^m} + \eta_f\right) - 2\left(\frac{G_{23}^f}{G^m}\eta_m - \eta_f\right)V^{f^3}\right]\right\}$$

$$C = -3V^f V^{m^2}\left(\frac{G_{23}^f}{G^m} - 1\right)\left(\frac{G_{23}^f}{G^m} + \eta_f\right) + \left[\frac{G_{23}^f}{G^m}\eta_m + \left(\frac{G_{23}^f}{G^m} - 1\right)V^f + 1\right]\left[\frac{G_{23}^f}{G^m} + \eta_f + \left(\frac{G_{23}^f}{G^m}\eta_m - \eta_f\right)V^{f^3}\right]$$

With

$$\eta_m = 3 - v_m \; ; \eta_f = 3 - v_{23}^f$$

$K_f = \frac{E_f}{2(1-2v_f)(1+v_f)}$ and $K_m = \frac{E_m}{2(1-2v_m)(1+v_m)}$ are the bulk modulus of the fiber and the matrix under longitudinal strain respectively .

$$v_{23} = \frac{K - m.G_{23}}{K + m.G_{23}}; \text{ with } m = 1 + 4K.\frac{v_{12}^2}{E_{11}}$$

K is the bulk modulus of the composite under longitudinal strain

$$K = \frac{K^m.(K^f + G^m).V^m + K^f.(K^m + G^m).V^f}{(K^f + G^m).V^m + (K^m + G^m).V^f}$$

$$E_{22} = 2.(1 + v_{23}).G_{23}$$

2.4. Homogenization models

2.4.1. Mori-Tanaka model (M-T)

The Mori-Tanaka model is initially developed by Mori and Tanaka [7]. This is a well-known model which is widely used for modeling different kinds of composite materials. This is an inclusion model, where fibers are simulated by inclusions embedded in a homogeneous medium. The Benveniste formulation [8] for the Mori-Tanaka model is given by:

$$C_{MT} = C_m + \left[V_f.\langle(C_f - C_m).A_{Eshelby}\rangle\right].\left[V_m.I + V_f.\langle A_{Eshelby}\rangle\right]^{-1}$$

With C_m and C_f are the stiffness matrices of the matrix phase and the reinforcement phase (inclusions) respectively. V_f and V_m are the volume fractions of the matrix phase and the reinforcement phase (inclusions) respectively. $A_{Eshelby}$ is the strain concentration tensor of the dilute solution presented by:

$$A_{\text{Eshelby}} = [I + E. C_m^{-1}. (C_f - C_m)]^{-1}$$

With E is the Eshelby tensor which depends on the shape of the inclusion and the Poisson's ratio of the matrix. More detailed information about the Eshelby tensor could be found in Mura [9]. The Eshelby tensor is then calculated for each inclusion along with the stiffness matrix.

2.4.2. Self-consistent model (S-C)

The self-consistent model has been proposed by Hill [10] and Budianski [11] to predict the elastic properties of composite materials reinforced by isotropic spherical particulates. Later the model was presented and used to predict the elastic properties of short fibers composites [12]. In this study the potential of the S-C model will be investigated when applied on UD composites with long fibers. The S-C model is an iterative model yielding the stiffness matrix as follows:

At the first iteration, fibers which represent the inclusions are supposed surrounded by an isotropic matrix, thus the S-C model is similar to the Eshelby dilute solution model. Then, at the second iteration, the inclusions are considered to be embedded in homogeneous medium which supposed to have the stiffness matrix similar to that of the composite calculated at the first iteration.

First iteration:

$$A_{\text{Eshelby}} = [I + E. C_m^{-1}. (C_f - C_m)]^{-1}$$

$$C_{sc} = C_m + [V_f. \langle (C_f - C_m). A_{\text{Eshelby}} \rangle]$$

Second iteration:

$$A_{\text{Eshelby}} = [I + E. C_{sc}^{-1}. (C_f - C_{sc})]^{-1}$$

$$C_{sc} = C_m + [V_f. \langle (C_f - C_m). A_{\text{Eshelby}} \rangle]$$

2.4.3. Bridging model

Recently, a new micromechanical model has been proposed by Huang et al. [13,14]. The model is developed to predict the stiffness and the strength of UD composites. The elastic properties by the bridging model is given as follows:

$$E_{11} = V_f. E_{11}^f + V_m. E_m$$

$$E_{22} = \frac{(V_f + V_m.a_{11})(V_f + V_m.a_{22})}{(V_f + V_m.a_{11})(V_f.S_{11}^f + V_m.a_{22}.S_{22}^m) + V_f.V_m(S_{21}^m - S_{21}^f)a_{12}}$$

$$V_{12} = V_f v_{11}^f + V_m v_m$$

$$G_{12} = \frac{(V_f + V_m.a_{66})G_{12}^f G_m}{V_f.G_m + V_m.a_{66}.G_{12}^f}$$

$$G_{23} = \frac{0.5(V_f + V_m \cdot a_{44})}{V_f(S_{22}^f - S_{23}^f) + V_m \cdot a_{44}(S_{22}^m - S_{23}^m)}$$

With a_{ij} are the components of the bridging matrix A, [13,14].

S_{ij}^f and and S_{ij}^m are the components of the compliance matrices of the fibers and the matrix respectively.

2.5. Numerical FE modeling

The numerical FE modeling is widely used in predicting the mechanical properties of composites. The numerical modeling is a reliable tool, but the time consumed on the geometrical dimensions definition and the corresponding calculation time, represent a major disadvantage against analytical models. Moreover there are many discussions and studies that deal with the appropriate boundary, symmetric and periodic conditions required to evaluate the elastic properties of UD composites. In this domain, a major work is done by S. Li [15]. It should be noticed that the numerical FE modeling require geometrical modeling or representation of the REV. while for UD composites, there are three types of fiber arrangements: square array, diamond array and hexagonal array (Figure 2).

In order to investigate the numerical FE modeling, the modeling of a quarter unit cell for a square array, diamond array and hexagonal array is conducted using Comsol Multiphysics software. A tetrahedral meshing is used. The resumed boundary conditions applied are given in (Table 1). Note that U, V and W are the displacements along 1, 2 and 3 directions respectively applied on the X+, X-, Y+, Y-, Z+ and Z- faces (with X faces are orthogonal to the fiber direction 1).

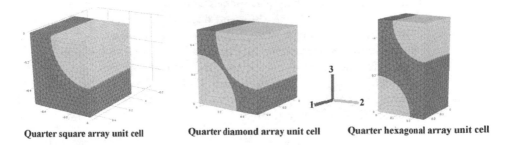

| Quarter square array unit cell | Quarter diamond array unit cell | Quarter hexagonal array unit cell |

Figure 2. Meshing of square, diamond and hexagonal array unit cells.

After applying boundary conditions and the displacement constant K, the corresponding engineering constants are calculated as follow, in terms of corresponding stresses and strains (σ_{11}, σ_{22}, τ_{12}, τ_{23}, ε_{11}, ε_{22}, Y_{12} and Y_{23}):

On the X+ face:

$E_{11} = \frac{\sigma_{11}}{\varepsilon_{11}}$, where σ_{11} and ε_{11} are calculated numerically on the X+ face

On the Y+ face:

$E_{22} = \frac{\sigma_{22}}{\varepsilon_{22}}$, where σ_{22} and ε_{22} are calculated numerically on the Y+ face

On the X+ face:

$G_{12} = \frac{\tau_{12}}{\gamma_{12}}$, where τ_{12} and γ_{12} are calculated numerically on the X+ face

On the Z+ face:

$G_{23} = \frac{\tau_{23}}{\gamma_{23}}$, where τ_{23} and γ_{23} are calculated numerically on the Z+ face

	X faces		Y faces		Z faces	
	X-	X+	Y-	Y+	Z-	Z+
E_{11} and ν_{12}	U = 0, V and W free	U = K, V and W free	V = 0, U and W free	U, V and W free	W = 0, U and V free	U, V and W free
E_{22} and ν_{23}	U = 0, V and W free	U, V and W free	V = 0, U and W free	V = K, U and W free	W = 0, U and V free	U, V and W free
G_{12}	V = W = 0	V = K W= 0	U = W = 0	U=W = 0	W = 0	W = 0
G_{23}	U = 0	U = 0	U = W = 0	U = W = 0	U = V = 0	U = 0 V= K

Table 1. Boundary conditions on the X, Y and Z faces of the quarter unit cell.

3. Comparative study, analysis and discussion

3.1. Results

In this section, a comparison of analytical models and numerical models with available experimental data is presented. Three different kinds of UD composites are taken as examples: Glass/epoxy composite [16], carbon/epoxy composite [14] and polyethylene/epoxy composite [17] (Table 2). The glass fibers are isotropic fibers while the carbon and the polyethylene fibers are transversely isotropic fibers. Knowing that the epoxy matrices are assumed isotropic, it's well noticed that for the polyethylene/epoxy, the Young's modulus of the epoxy is higher than that transversal modulus of the fibers, which represent an important case to be investigated.

Fibers	E_{11}^f (GPa)	E_{22}^f (GPa)	G_{12}^f (GPa)	ν_{12}^f	ν_{23}^f
E-Glass [16]	73.1	73.1	29.95	0.22	0.22
Carbon [14]	232	15	24	0.279	0.49
Polyethylene [17]	60.4	4.68	1.65	0.38	0.55
Matrix	E^m (GPa)		G^m (GPa)	ν^m	
Epoxy resin [16]	3.45	3.45	1.28	0.35	0.35
Epoxy [14]	5.35	5.35	1.97	0.354	0.354
Epoxy [17]	5.5	5.5	1.28	0.37	0.37

Table 2. Elastic properties of the fibers and epoxy matrices.

3.1.1. Longitudinal Young's modulus E_{11}

For the longitudinal Young's modulus E_{11}, obtained analytical and numerical results are compared to those available experimental data for carbon/epoxy and polyethylene/epoxy UD composites in terms of the fiber volume fraction Vf. Investigated analytical models belong to the ROM, the Elasticity approach model (EAM), M-T and S-C models. Please note that ROM, MROM, Chamis, Halpin-Tsai and Bridging models share the same formulation for E_{11}.

It's well noticed that the predicted results for all investigated models are in good agreement with the experimental data for both composites with different V^f (Figure 3 and 4).

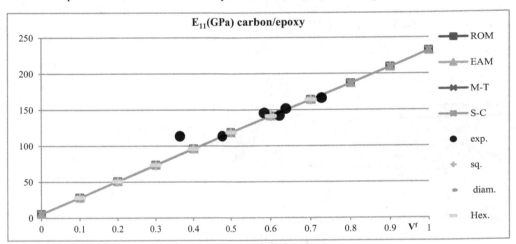

Figure 3. Predicted analytical, numerical and experimental results for E_{11} in terms of V^f.

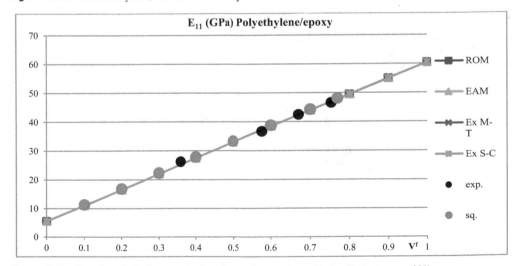

Figure 4. Predicted analytical, numerical and experimental results for E_{11} in terms of V^f.

3.1.2. *Transversal Young's modulus E_{22}*

The prediction of the transversal Young's modulus and in contrast with the longitudinal modulus presents a real challenge for the researchers. Thus, many analytical models are proposed belonged to different micromechanics approach. In addition, the potential of the FE element modeling is investigated. Predicted results of different analytical and numerical models for three UD composites are presented in figures (5,6 and 7)

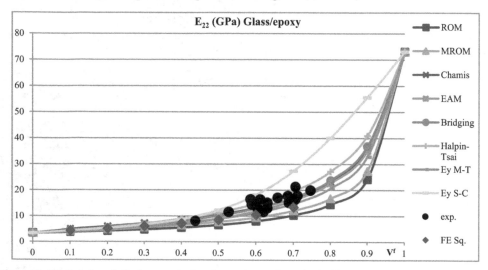

Figure 5. Predicted analytical, numerical and experimental results for E_{22} in terms of V^f.

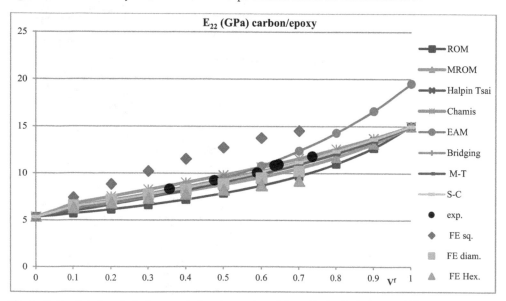

Figure 6. Predicted analytical, numerical and experimental results for E_{22} in terms of V^f.

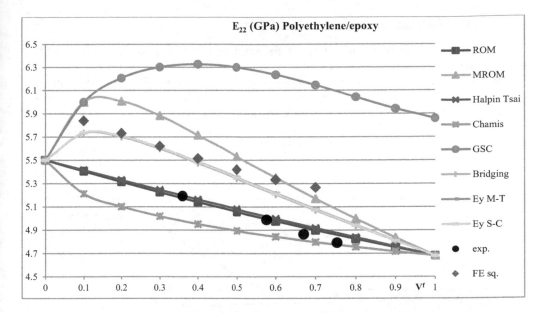

Figure 7. Predicted analytical, numerical and experimental results for E₂₂ in terms of Vᶠ.

It's shown that for the glass/epoxy composite, the S-C model overestimates the experimental results, while the ROM and MROM models underestimate it. Other analytical models, especially the Chamis, Bridging and EAM models yield results that correlate well with the available experimental data for different values of Vf. Moreover, it's noticed that the FE (Square array), the Halpin-Tsai and the M-T models gives good predictions. Concerning composites reinforced with transversely isotropic fibers, it's well remarked that the EAM model well overestimates E₂₂ especially with the polyethylene/epoxy composite. The ROM underestimates the experimental results, while other analytical models, in addition to the numerical FE (diamond array) model, yield very good predictions for the carbon/epoxy composite. However, with the polyethylene/epoxy, it's noticed that only the results obtained from the ROM and the Halpin-Tsai models correlate well with the experimental data, while the Chamis model shows a good agreement with Vf higher than 0.6.

3.1.3. Longitudinal shear modulus G₁₂

Experimental results for two UD composites are used to be compared with. Figures 8 show clearly the MROM, EAM, Halpin-Tsai, Chamis, bridging analytical models, in addition to all numerical FE models yield very good results for the carbon/epoxy composite. However, it's remarked that the inclusion models, the M-T and S-C models, overestimate the longitudinal shear modulus. Concerning the polyethylene/epoxy composite, only results obtained results from the MROM and Chamis models agree well with the available experimental data (Figure 9).

3.1.4. Transversal shear modulus G_{23}

For the transversal shear modulus G_{23}, it's shown from Figure 10 and 11, that the bridging model yields the best results. In addition, it's remarked that the EAM, Chamis yield reasonable predictions underestimating the experimental data, while the M-T and S-C models overestimate it. Concerning the numerical modeling, predicted results always overestimate the available experimental results for the two composites.

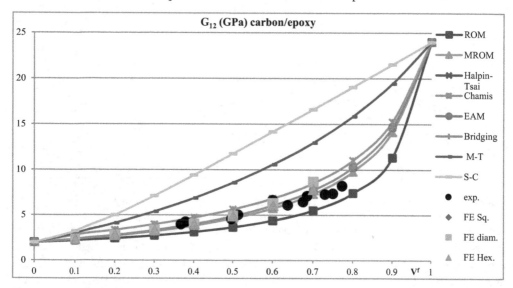

Figure 8. Predicted analytical, numerical and experimental results for G_{12} in terms of V^f.

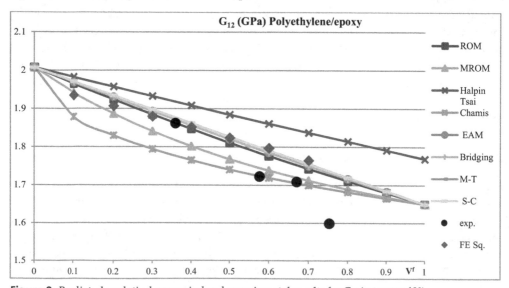

Figure 9. Predicted analytical, numerical and experimental results for G_{12} in terms of V^f.

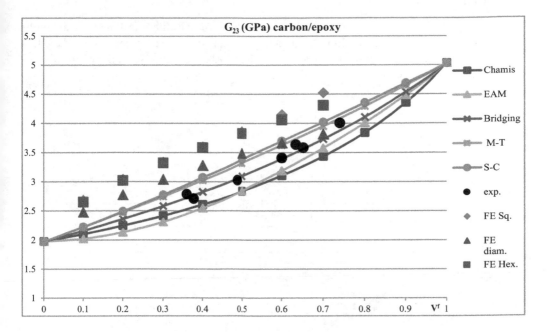

Figure 10. Predicted analytical, numerical and experimental results for G₂₃ in terms of Vᶠ.

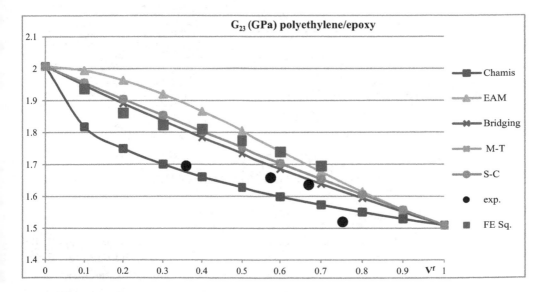

Figure 11. Predicted analytical, numerical and experimental results for G₂₃ in terms of Vᶠ.

3.1.5. Major Poisson's ratio v_{12}

Concerning the Poisson's ratios, the obtained results of the analytical models are only compared to those numerical due the missing of experimental data for the studied UD composites. Figure 12 shows that for the major Poisson's ratio v_{12}, all analytical and numerical models correlate well with each other.

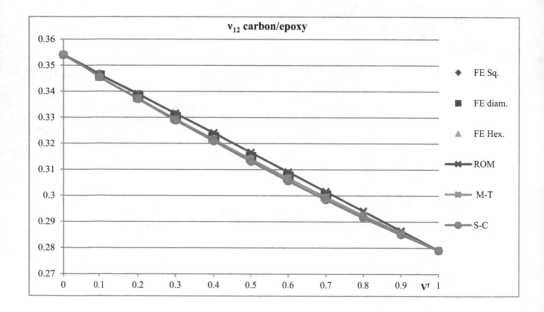

Figure 12. Predicted analytical and numerical results for v_{12} in terms of V^f.

3.2. Analysis and discussion

In this section, an analysis of the predicted results for each model is presented apart. It's shown from the above results that for the phenomenological models, the Voigt and Reuss models, represented by the ROM model, show very good predictions for the longitudinal Young's modulus E_{11} and major Poisson's ratio v_{12}. However, with for the transversal Young's modulus E_{22} the ROM model always underestimates the experimental results except for the polyethylene/epoxy case where it's well agree with the available experimental data. Concerning the longitudinal shear modulus G_{12}, the ROM model didn't yield good prediction for both studied cases the carbon/epoxy and the polyethylene/epoxy composites.

As known the semi-empirical models have been emerged and proposed in order to correct the predictions of the ROM model for the transversal Young's and longitudinal shear moduli. While the investigated models share the same formulations for E_{11} and v_{12} with ROM model, the corrections made for E_{22} and G_{12} prove to be effective. It's shown that the Chamis model yields very good results for all studied cases, while the MROM and Halpin-Tsai the models only suffer with the special case of the polyethylene/epoxy with the E_{22} and G_{12} respectively.

Concerning the elasticity approach models, the proposed formulation of the E_{11} yields similar results for that proposed by the ROM model. While for the transversal Young's modulus E_{22}, it's clearly noticed that with isotropic fibers, the model results correlate well with those experimental, while with the case of transversely isotropic fibers, reasonable predictions are shown for the carbon/epoxy case. However, for the polyethylene case the model well overestimates the experimental results. The reason could be conducted to that EAM models are initially proposed to deal with UD composites reinforced with isotropic fibers. For the longitudinal shear modulus G_{12}, the elastic solution formulation agrees well with the experimental data. Concerning the transversal shear modulus, the predictions made by the generalized self-consistent model of the Christensen model [6], which is developed to enhance the predictions of this elastic property, always overestimates the experimental data.

In this study, the potential of the homogenization models is investigated. The inclusion models, the M-T and the S-C models, and the bridging model, yield good prediction for both longitudinal Young's modulus and major Poisson's ratio. However, for the transversal Young's modulus E_{22}, reasonable agreement is shown for the glass/epoxy and carbon/epoxy cases, except with the self-consistent model which overestimates the experimental data for high V^f. While for the case of the polyethylene/epoxy, all three models yield almost the same results and overestimate the compared experimental data while agree with FE modeling results. The same problem is shown with the prediction of the shear moduli, where for the polyethylene/epoxy case, the models belonged to the homogenization approach give the same results overestimating the experimental data. While with the carbon/epoxy case, it's noticed that the bridging model predicts better the shear moduli, while the M-T and S-C models well overestimate the experimental data especially for the G_{12}.

Concerning the numerical modeling, it's well noticed that there are different predicted results for different arrays. It's also remarked, that the FE numerical modeling didn't yield better results than the analytical models, except for the longitudinal Young's modulus and major Poisson's ratio where all predicted results from numerical and analytical models correlate well with available experimental data.

4. Conclusion

In this study, the evaluated results, for the elastic properties, of most known analytical micromechanical models, as well as FE modeling methods, are compared to available experimental data for three different UD composites: Glass/epoxy, carbon/epoxy and polyethylene/epoxy. It should be noticed that the studied cases cover different kinds of reinforced composites by isotropic fibers (glass) and transversely isotropic fibers (carbon and polyethylene). In addition, the polyethylene/epoxy presents an interesting case study, where the matrix is stiffer than the fibers in the transvers direction.

The analyses of the compared results show clearly that all analytical and numerical models show a very good agreement for the longitudinal Young's modulus E_{11} and major Poisson's ratio v_{12}. However, the other moduli, the transversal Young's modulus E_{22}, longitudinal shear modulus G_{12} and the transversal shear modulus G_{23}, represent the main challenge for the researchers. It's shown that analytical micromechanical models belonged to the semi-empirical models, especially the Chamis model, predict well these elastic properties. Moreover, the bridging model proves to be a reliable model when predicting the elastic properties of carbon/epoxy composite. It's noticed that almost all models suffer with the prediction of elastic properties for the polyethylene/epoxy composite. However, models belonged to the elasticity approach and inclusion approach (M-T and S-C models) show inconsistency in predicting the elastic properties of studied UD composites. Numerical models, based on the FE method, show that using different fibers arrangements will lead to different predicted results. Moreover, the FE didn't prove that it could be more accurate than some simple and straightforward analytical model. As a conclusion from this study, the Chamis model and the bridging model could be considered as the most complete models which could give quite accurate estimations for all five independent elastic properties. Noting that the corrections proposed by the Halpin-Tsai model, prove that it well enhance the prediction of the transversal Young's modulus E_{22}.

Author details

Rafic Younes[*]
LISV, University of Versailles Saint-Quentin, Versailles, France
Faculty of Engineering, Lebanese University, Rafic Hariri campus, Beirut, Lebanon

Ali Hallal
LISV, University of Versailles Saint-Quentin, Versailles, France
L3M2S, Lebanese University, Rafic Hariri campus, Beirut, Lebanon

Farouk Fardoun
L3M2S, Lebanese University, Rafic Hariri campus, Beirut, Lebanon

---

[*] Corresponding Author

Fadi Hajj Chehade
L3M2S, Lebanese University, Rafic Hariri campus, Beirut, Lebanon

5. References

[1] Voigt W. Uber die Beziehung zwischen den beiden Elastizitatskonstanten Isotroper Korper. Wied. Ann, 38 (1889) 573-587.

[2] Reuss A. Berechnung der Fliessgrense von Mischkristallen auf Grund der Plastizitätsbedingung für Einkristalle. Zeitschrift Angewandte Mathematik und Mechanik, 9 (1929) 49-58.

[3] Halpin JC, Kardos JL. The Halpin-Tsai equations: A review. Polymer Engineering and Science, May, 1976, Vol. 16, No. 5.

[4] Chamis CC. Mechanics of composite materials: past, present, and future. J Compos Technol Res ASTM 1989;11:3–14.

[5] Hashin Z, Rosen BW. The elastic moduli of fiber reinforced materials. Journal of Applied Mechanics, Trans ASME, 31 (1964). 223-232.

[6] Christensen RM. A critical evaluation for a class of micromechanics models. Journal of Mechanics and Physics of Solids 1990; 38(3):379–404.

[7] Mori T, Tanaka K. Average stress in matrix and average elastic energy of materials with misfitting inclusions, Acta Mettall. 21, 1973, 571-574.

[8] Benveniste Y. A new approach to the application of Mori-Tanaka's theory in composite materials. Mechanics of Materials, 6, 1987, 147-157.

[9] Mura T. Micromechanics of Defects in Solids, 2nd edn. Martinus Nijhof Publishers, Dordrecht, 1987.

[10] Hill R. Theory of mechanical properties of fibre-strengthen materials-III. Self-consistent model. Journal of Mechanics and Physics of Solids 1965; 13: 189-198.

[11] Budiansky B. On the elastic moduli of some heterogeneous materials, J. Mech. Phys. Solids 13, 1965, 223-227.

[12] Chou TW, Nomura S, Taya M. A self-consistent approach to the elastic stiffness of short-fiber composites, J. Compos Mater, 14, 1980, 178-188.

[13] Huang ZM. Simulation of the mechanical properties of fibrous composites by the bridging micromechanics model. Composites: Part A 32 (2001) 143–172.

[14] Huang ZM. Micromechanical prediction of ultimate strength of transversely isotropic fibrous composites. International Journal of Solids and Structures 38 (2001) 4147-4172.

[15] Li S. Boundary conditions for unit cells from periodic microstructures and their implications. Composites Science and Technology 68 (2008) 1962–1974.

[16] Shan HZ, Chou TW. Transverse elastic moduli of unidirectional fiber composites with fiber/matrix interfacial debonding. Composites Science and Technology 53 (1995) 383-391.

[17] Wilczynski AP, Lewinski J. Predecting the properties of unidirectional fibrous composites with monotropic reinforcement. Composite Science and Technology 55 (1995) 139-143

Metal and Ceramic Matrix Composites

Manufacturing and Properties of Quartz (SiO₂) Particulate Reinforced Al-11.8%Si Matrix Composites

M. Sayuti, S. Sulaiman, T.R. Vijayaram, B.T.H.T Baharudin and M.K.A. Arifin

Additional information is available at the end of the chapter

1. Introduction

Metal matrix composites (MMC) are a class of composites that contains an element or alloy matrix in which a second phase is fixed firmly deeply and distributed evenly to achieve the required property improvement. The property of the composite varies based on the size, shape and amount of the second phase (Sayuti et al., 2010; Sulaiman et al., 2008). Discontinuously reinforced metal matrix composites, the other name for particulate reinforced composites, constitute 5 – 20 % of the new advanced materials (Gay et al., 2003). The mechanical properties of the processed composites are greatly influenced by their microstructure. An increased stiffness, yield strength and ultimate tensile strength are generally achieved by increasing the weight fraction of the reinforcement phase in the matrix. Inspite of these advantages, the usage of particulate reinforced MMCs as structural components in some applications is limited due to low ductility (Rizkalla and Abdulwahed, 1996). Owing to this and to overcome the draw-backs, a detailed investigation on the strengthening mechanism of composites has been carried out by composite experts (Humphreys, 1987). They have found that the particle size and its weight fraction in metal matrix composites influences the generation of dislocations due to thermal mismatch. The effect is also influenced by the developed residual and internal stresses too. The researchers have predicted that the dislocation density is directly proportional to the weight fraction and due to the amount of thermal mismatch. As a result, the strengthening effect is proportional to the square root of the dislocation density. This effect would be significant for fine particles and for higher weight fractions. The MMCs yield improved physical and mechanical properties and these outstanding benefits are due to the combined metallic and ceramic properties (Hashim et al., 2002). Though there are various types of MMCs, particulate-reinforced composites are the most versatile and economical (Sayuti, Sulaiman, Vijayaram, et al., 2011; Sayuti, Suraya, et al., 2011).

In the past 40 years, the researchers and design experts have perceived their research to emphasis on finding lightweight, environmental friendly, low-cost, high quality, and good performance materials (Feest, 1986). In accordance with this trend, MMCs have been attracting growing interest among researchers and industrialists. The attributes of MMCs include alterations in mechanical behavior (e.g., tensile and compressive properties, creep, notch resistance, and tribology) and physical properties (e.g., intermediate density, thermal expansion, and thermal diffusivity) a change, primarily induced by the reinforced filler phase (Sayuti, et al., 2011). Even though MMCs posses various advantages, they still have limitations of thermal fatigue, thermo-chemical compatibility, and posses lower transverse creep resistance. In order to overcome these limitations, fabrication of discontinuously reinforced Al-based MMCs was carried out by standard metallurgical processing methods such as powder metallurgy, direct casting, rolling, forging and extrusion. Subsequently, the products were shaped, machined and drilled by using conventional machining processes. Consequently, the MMCs would be available in suitable quantities with desirable properties, particularly for automotive applications (Sharma et al., 1997).

In general, composite materials posses good mechanical and thermal properties, sustainable over a wide range of temperatures (Vijayaram et al., 2006). The desirable factors such as property requirements, cost factor considerations and future application prospects would decide the choice of the processing method (Kaczmar et al., 2000). In practice, composite materials with a metal or an alloy matrix are fabricated either by casting or by powder metallurgy methods (Fridlyander, 1995). They are considered as potential material candidates for a wide variety of structural applications in the transportation, automobile and sport goods manufacturing industries due to the superior range of mechanical properties they exhibit (Hashim et al., 1999). MMCs represent a new generation of engineering materials in which a strong ceramic reinforcement is incorporated into a metal matrix to improve its properties such as specific strength, specific stiffness, wear resistance, corrosion resistance and elastic modulus (Baker et al., 1987; Chambers et al., 1996; Kok, 2005). As a virtue of their structure and bonding between the matrix and the reinforcement, MMCs combine metallic properties of matrix alloys (ductility and toughness) with ceramic properties of reinforcements (high strength and high modulus), therein leading to greater strength in shear and compression as well as higher service-temperature capabilities (Huda et al., 1993). Thus, they have scientific, technological and commercial significance. MMCs, because of their improved properties, are being used extensively for high performance applications such as in aircraft engines especially in the last decade. Recently, they also find application in automotive sectors (Surappa, 2003; Therén and Lundin, 1990).

Aluminum oxide (Al_2O_3) and silicon carbide (SiC) powders in the form of fibers and particulates are commonly used as reinforcements in MMCs. In the automotive and aircraft industries for example, production of engine pistons and cylinder heads, the tribological properties of the materials used are considered crucial. Hence, Aluminum oxide and silicon carbide reinforced aluminum alloy matrix composites are applied in these fields (Prasad and

Asthana, 2004). Due to their high demand, the development of aluminum matrix composites is receiving considerable emphasis in modern application. Research reports ascertain that the incorporation of hard second phase particles in the alloy matrices to produce MMCs is beneficial and economical due to its high specific strength and corrosion resistance properties (Kok, 2005). Therefore, MMCs are those materials that have higher potential for a large range of engineering applications.

2. Metal Matrix Composites (MMCs)

Metal matrix composites are a family of new materials which are attracting considerable industrial interest and investment worldwide. They are defined as materials whose microstructures compromise a continuous metallic phase (the matrix) into which a second phase, or phases, have been artificially introduced. This is in contrast to conventional alloys whose microstructures are produced during processing by naturally occurring phase transformations (Feest, 1986). Metal matrix composites are distinguished from the more extensively developed resin matrix composites by virtue of their metallic nature in terms of physical and mechanical properties and by their ability to lend themselves to conventional metallurgical processing operations. Electrical conductivity, thermal conductivity and non-inflammability, matrix shear strength, ductility (providing a crack blunting mechanism) and abrasion resistance, ability to be coated, joined, formed and heat treated are some of the properties that differentiate metal matrix composites from resin matrix composites. MMCs are a class of advanced materials which have been developed for weight-critical applications in the aerospace industry. Discontinuously reinforced aluminum composites, composed of high strength aluminum alloys reinforced with silicon carbide particles or whiskers, are a subclass of MMCs. Their combination of superior properties and fabricability makes them attractive candidates for many structural components requiring high stiffness, high strength and low weight. Since the reinforcement is discontinuous, discontinuously reinforced composites can be made with properties that are isotropic in three dimensions or in a plane. Conventional secondary fabrication methods can be used to produce a wide range of composites products, making them relatively inexpensive compared to the other advanced composites reinforced with continuous filaments. The benefit of using composite materials and the cause of their increasing adoption is to be looked for in the advantage of attaining property combinations that can result in a number of service benefits. Among these are increased strength, decreased weight, higher service temperature, improved wear resistance and higher elastic module. The main advantage of composites lies in the tailorability of their mechanical and physical properties to meet specific design criteria.   Composite materials are continuously displacing traditional engineering materials because of their advantages of high stiffness and strength over homogeneous material formulations. The type, shape and spatial arrangement of the reinforcing phase in metal matrix composites are key parameters in determining their mechanical behavior. The hard ceramic component which increases the mechanical characteristics of metal matrix composites causes quick wear and premature tool failure in the machining operations. Metal matrix composites have been investigated since the early 1960s with the impetus at that time, being the high potential structural properties that would be achievable with materials engineered to specific applications (Mortensen et al., 1989).

In the processing of metal matrix composites, one of the subjects of interest is to choose a suitable matrix and a reinforcement material (Ashby and Jones, 1980). In some cases, chemical reactions that occur at the interface between the matrix and its reinforcement materials have been considered harmful to the final mechanical properties and are usually avoided. Sometimes, the interfacial reactions are intentionally induced, because the new layer formed at the interface acts as a strong bond between the phases (Gregolin et al., 2002).

During the production of metal matrix composites, several oxides have been used as reinforcements, in the form of particulates, fibers or as whiskers (Zhu and Iizuka, 2003). For example, alumina, zirconium oxide and thorium oxide particulates are used as reinforcements in aluminum, magnesium and other metallic matrices (Upadhyaya, 1990). Very few researchers have reported on the use of quartz as a secondary phase reinforcement particulate in an aluminum or aluminum alloy matrix, due to its aggressive reactivity between these materials (Sahin, 2003). Preliminary studies showed that the contact between molten aluminum and silica-based ceramic particulates have destroyed completely the second phase microstructure, due to the reduction reaction which provokes the infiltration of liquid metal phase into the ceramic material (Mazumdar, 2002). Previous works carried out by using continuous silica fibers as reinforcement phases in aluminum matrix showed that even at temperatures nearer to 400 °C, silica and aluminum can react and produce a transformed layer on the original fiber surface as a result of solid diffusion between the phases and due to the aluminum-silicon liquid phase formation (Seah et al., 2003). The organizations and companies that are very active in the usage of MMCs in Canada and United States include the following (Rohatgi, 1993):

1. Aluminium Company of Canada, Dural Corporation, Kaiser Aluminium, Alcoa, American Matrix, Lanxide, American Refractory Corporation
2. Northrup Corporation, McDonald Douglas, Allied Signal, Advanced Composite Materials Corporation, Textron Specialty Materials
3. DWA Associates, MCI Corporation, Novamet
4. Martin Marietta Aerospace, Oakridge National Laboratory , North American Rockwell, General Dynamics Corporation, Lockheed Aeronautical Systems
5. Dupont, General Motors Corporation, Ford Motor Company, Chrysler Corporation, Boeing Aerospace Company, General Electric , Westinghouse
6. Wright Patterson Air Force Base, (Dayton, Ohio), and
7. Naval Surface Warefare Centre, (Silver Spring, Maryland)

India also has substantial activity in PM and cast MMCs. It has had world class R&D in cast aluminium particulate composites which was sought even by western countries.

2.1. Classification of composites

Among the major developments in materials in recent years are composite materials. In fact, composites are now one of the most important classes of engineered materials, because they offer several outstanding properties as compared to conventional materials. The matrix material in a composite may be ceramic based, polymer or metal. Depending on the matrix, composite materials are classified as follows:

Metal matrix composites (MMCs)
Polymer matrix composites (PMCs)
Ceramic matrix composites (CMCs)

Majority of the composites used commercially are polymer-based matrices. However, metal matrix composites and ceramic matrix composites are attracting great interest in high temperature applications (Feest, 1986). Another class of composite material is based on the cement matrix. Because of their importance in civil engineering structures, considerable effort is being made to develop cement matrix composites with high resistance to cracking (Schey, 2000). Metal matrix composites (MMCs) are composites with a metal or alloy matrix. It has resistance to elevated temperatures, higher elastic modulus, ductility and higher toughness. The limitations are higher density and greater difficulty in processing parts. Matrix materials used in these composites are usually aluminum, magnesium, aluminum-lithium, titanium, copper and super alloys. Fiber materials used in MMCs are aluminum oxide, graphite, titanium carbide, silicon carbide, boron, tungsten and molybdenum. The tensile strengths of non metallic fibers range between 2000 MPa to 3000 MPa, with elastic modulus being in the range of 200 GPa to 400 GPa. Because of their lightweight, high specific stiffness and high thermal conductivity, boron fibers in an aluminum matrix have been used for structural tubular supports in the space shuttle orbiter. Metal matrix composites having silicon carbide fibers and a titanium matrix, are being used for the skin, stiffeners, beams and frames of the hypersonic aircrafts under development. Other applications are in bicycle frames and sporting goods (Wang et al., 2006). Graphite fibers reinforced in aluminum and magnesium matrices are applied in satellites, missiles and in helicopter structures. Lead matrix composites having graphite fibers are used to make storage-battery plates. Graphite fibers embedded in copper matrix are used to fabricate electrical contacts and bearings. Boron fibers in aluminum are used as compressor blades and structural supports. The same fibers in magnesium are used to make antenna structures. Titanium-boron fiber composites are used as jet-engine fan blades. Molybdenum and tungsten fibers are dispersed in cobalt-base super alloy matrices to make high temperature engine components. Squeeze cast MMCs generally have much better reinforcement distribution than compocast materials. This is due to the fact that a ceramic perform which is used to contain the desired weight fraction of reinforcement rigidly attached to one another so that movement is inhibited. Consequently, clumping and dendritic segregation are eliminated. Porosity is also minimized, since pressure is used to force the metal into interfiber channels, displacing the gases. Grain size and shape can vary throughout the infiltrated preform because of heat flow patterns. Secondary phases typically form at the fiber-matrix interface, since the lower freezing solute-rich regions diffuse toward the fiber ahead of the solidifying matrix (Surappa, 2003).

2.2. Significance of composites

Composites technology and science requires interaction of various disciplines such as structural analysis and design, mechanics of materials, materials science and process engineering. The tasks of composites research are to investigate the basic characteristics of the constituents and composite materials, develop effective and efficient fabrication procedures, optimize the material for service conditions and understanding their effect on material properties and to determine material properties and predict the structural behavior by analytical procedures and hence to develop effective experimental techniques for material characterization, failure analysis and stress analysis (Daniel and Ishai, 1994). An important task is the non-destructive evaluation of material integrity, durability assessment, structural reliability, flaw criticality and life prediction. The structural designs and systems capable of operating at elevated temperatures has spurred intensive research in high temperature composites, such as ceramic/matrix, metal/ceramic and carbon/carbon composites. The utilization of conventional and new composite materials is intimately related to the development of fabrication methods. The manufacturing process is one of the most important stages in controlling the properties and ensuring the quality of the finished product. The technology of composites, although still developing, has reached a state of maturity. Nevertheless, prospects for the future are bright for a variety of reasons. Newer high volume applications, such as in the automotive industry, will expand the use of composites greatly.

2.3. Matrix

Matrix is the percolating alloy/metal/polymer/plastic/resin/ceramic forming the constituent of a composite in which other constituents are embedded. If the matrix is a metal, then it is called as a metal matrix and consecutively polymer matrix, if the matrix is a polymer and so on. In composites, the matrix or matrices have two important functions (Weeton et al., 1988). Firstly, it holds the reinforcement phase in the place. Then, under an applied force, it deforms and distributes the stress to the reinforcement constituents. Sometimes the matrix itself is a key strengthening element. This occurs in certain metal matrix composites. In other cases, a matrix may have to stand up to heat and cold. It may conduct or resist electricity, keep out moisture, or protect against corrosion. It may be chosen for its weight, ease of handling, or any of many other applications. Any solid that can be processed to embed and adherently grip a reinforcing phase is a potential matrix material.

In a composite, matrix is an important phase, which is defined as a continuous one. The important function of a matrix is to hold the reinforcement phase in its embedded place, which act as stress transfer points between the reinforcement and matrix and protect the reinforcement from adverse conditions (Clyne, 1996). It influences the mechanical properties, shear modulus and shear strength and its processing characteristics. Reinforcement phase is the principal load-carrying member in a composite. Therefore, the orientation, of the reinforcement phase decides the properties of the composite.

2.4. Reinforcing phase / Materials

Reinforcement materials must be available in sufficient quantities and at an economical rate. Recent researches are directed towards a wider variety of reinforcements for the range of matrix materials being considered, since different reinforcement types and shapes have specific advantages in different matrices (Basavarajappa et al., 2004). It is to be noted that the composite properties depend not only on the properties of the constituents, but also on the chemical interaction between them and on the difference in their thermal expansion coefficients, which both depend on the processing route. In high temperature composites, the problem is more complicated due to enhanced chemical reactions and phase instability at both processing and application temperatures. Reinforcement phases in MMCs are embedded in the form of continuous reinforcement or discontinuous reinforcement in the matrix material. The reinforcing phase may be a particulate or a fiber, continuous type or discontinuous type. Some of the important particulates normally reinforced in composite materials are titanium carbide, tungsten carbide, silicon nitride, aluminum silicate, quartz, silicon carbide, graphite, fly ash, alumina, glass fibers, titanium boride etc. The reinforcement second phase material is selected depending on the application during the processing of composites (Clyne, 1996). The reinforcement phase is in the form of particulates and fibers generally. The size of the particulate is expressed in microns, micrometer. However, the discontinuous fiber is defined by a term called as 'Aspect Ratio'. It is expressed as the ratio of length to the diameter of the fiber. To improve the wettabilty with the liquid alloy or metal matrix material, the reinforcement phase is always preheated (Adams et al., 2003).

2.5. Factors affecting reinforcement

The interface between the matrix and the reinforcement plays an important role for deciding and explaining the toughening mechanism in the metal matrix composites. The interface between the matrix and the reinforcement should be organized in such a way that the bond in between the interface should not be either strong or weak (Singh et al., 2001).

2.6. Matrix interface / Interphase of matrices

Interfaces are considered particularly important in the mechanical behavior of MMCs since they control the load transfer between the matrix and the reinforcement. Their nature depends on the matrix composition, the nature of the reinforcement, the fabrication method and the thermal treatments of the composite. For particular matrix/reinforcement associations and especially with liquid processing routes, reactions can occur which change the composition of the matrix and lead to interfacial reaction products, thus changing the mechanical behavior of the composites. The interfacial phenomena in MMCs have been surveyed by several authors. Considering physical and chemical properties of both the matrix and the reinforcing material, the actual strength and toughness desired for the final MMCs, a compromise has to be achieved balancing often several conflicting requirements. A weak interface will lead to crack propagation following the interface, while a strong matrix associated with a strong interface will reveal cracks across both the matrix and the reinforcements. If however the matrix is weak in comparison with the interface and the particle strength, the failure will propagate through the matrix itself. The wettability of the reinforcement material by the liquid metallic matrix plays a major role in the bond formation. It mainly depends on heat of formation, electronic structure of the reinforcement and the molten metal temperature, time, atmosphere, roughness and crystallography of the reinforcement. Similarity between metallic bond and covalent bond is reflected in some metal, like titanium carbide and zirconium carbide which are more easily wetted than strong ionic bonds found in ceramics such as alumina that remains poorly wetted. Surface roughness of the reinforced material improves the mechanical interlocking at the interface, though the contribution of the resulting interfacial shear strength is secondary compared to chemical bonding. Large differences in thermal expansion coefficient between the matrix and the reinforcement should be avoided as they can include internal matrix stresses and ultimately give rise to interfacial failures. From a purely thermo dynamical point of view, a comparison of free enthalpy of formation at various temperatures shows that many metals in the liquid state are reactive toward the reinforcing materials in particular oxides or carbides. Though thermodynamically favored, some reactions are however not observed and practically the kinetics of these reactions has to be considered in conjunction with thermodynamic data in order to evaluate the real potential of the reactions. The consequences of such interfacial reactions are the chemical degradation of the reinforcing material associated with a decrease of its mechanical properties, the formation of brittle reaction products at the interface, as well as the release of elements initially part of the reinforcing material toward the matrix may generate inopportune metallurgical phases at the vicinity of the reinforcing materials. Moreover in the case of alloyed matrices, the selective reactivity and depletion of given elements from the alloy can generate compositional gradients in the matrix and may therefore alter its properties close to the interface. Though a moderate reaction may improve the composite bonding, extended reactions usually ruin the reinforcing material. The relation between interfacial reactions and interface strength depends on the materials. The elaboration of MMC requires often a very short solidification time to avoid excess interfacial reaction. During the cooling process, differences in thermal capacity and thermal conductivity between the reinforcing material and the matrix induce localized temperature gradients. Solidification of the metallic matrix is believed to be generally a directional outward process, starting from the inside of the metallic matrix while ending at the reinforcing material surface. Finally, the processing type and the parameters have to be selected and adjusted to a particular MMC system. Metals are generally more reactive in the liquid rather than in the solid state. Consequently, shorter processing time, that is, short contact time between the liquid metal and the reinforcement can limit the extent of interfacial reactions. The study of reinforcement and matrix bonding is important in composite matrix structure, which has been described by Gregolin (2002). While the load is acting on the composite, it has been distributed to the matrix and the reinforcement phase through the matrix interface. The reinforcement is effective in strengthening the matrix only if a strong interfacial bond exists between them. The interfacial properties also influence the resistance to crack propagation in a composite and therefore its fracture toughness (Dusza and Sajgalik, 1995). The two most important energy-absorbing failure mechanisms in a composite are debonding and particle pull-out at the particle matrix interface. If the interface between the matrix and reinforcement debonds, then the crack propagation is interrupted by the debonding process and instead of moving through the particle, the crack moves along the particle surface allowing the particle to carry a higher load (El-Mahallawy and Taha, 1993).

2.7. Physical phenomena of wettability and application

Wettability is defined as the extent to which a liquid will spread over a solid surface. Interfacial bonding is due to the adhesion between the reinforcement phase and the matrix. For adhesion to occur during the manufacturing of a composite, the reinforcement and the matrix must be brought into an intimate contact. During a stage in composite manufacture, the matrix is often in a condition where it is capable of flowing towards the reinforcement and this behavior approximates to that of the flow of a liquid. A key concept in this contact is wettability. Once the matrix wets the reinforcement particle, and thus the matrix being in intimate contact with the reinforcement, causes the bonding to occur (Hashim et al., 2001; Oh et al., 1987). Different types of bonding will occur and the type of bonding varies from system to system and it entirely depends on the details such as the presence of surface contaminants. The different types of bonding observed are mechanical bonding, electrostatic bonding, chemical bonding, and inter diffusion bonding (Burr et al., 1995). The bonding strength can be measured by conducting the tests like single particle test, bulk specimen test, and micro-indention test (Dusza and Sajgalik, 1995).

Poor wettability of most ceramic particulates with the molten metals is a major barrier to processing of these particulate reinforced MMCs by liquid metallurgy route. The characterization and enhancement of wettability is therefore, of central importance to successful composite processing (Asthana and Rohatgi, 1993). Wettability is shown in the Figure 1 below and it is customarily represented in terms of a contact angle defined from the Young-Dupre equation which is expressed as follows:

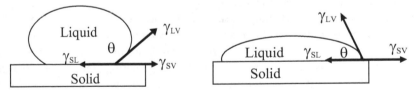

Figure 1. A sessile drop to the left is an example of poor wetting (θ>90°) and the sessile drop to the right is an example of good wetting (θ<90°) (Rajan et al., 1998).

$$\gamma_{lv} \cos \theta = \gamma_{sv} - \gamma_{sl} \tag{1}$$

Where γ_{SV} = Solid/Vapor surface energy, γ_{SL} = Solid/Liquid surface energy and γ_{LV} = Liquid/Vapor surface energy.

The wetting behavior of a liquid on a solid can be characterized by the wetting or contact angle that is formed between the liquid and the solid substrate. A "sessile drop" is a continuous drop of liquid on a flat, solid surface under steady-state conditions. To neglect the effects of gravity, the gravitational forces should be small compared to the surface tension of the drop. If this condition is satisfied, the drop will approach a hemispherical shape which represents its smallest area and lowest surface free energy. The sessile drop is placed on the solid substrate and the angle between the solid surface and the tangent to the liquid surface at the contact point is measured. This is known as the contact angle or wetting angle. The contact angle can vary between 0 and 180° and is a measure of the extent of wetting. The conditions of good wetting ($\theta < 90°$) and partial wetting ($\theta > 90°$) are illustrated in Figure 1. Complete wetting (also referred to as spreading) is obtained at an angle of 0° and complete non-wetting occurs at an angle of 180°. The contact angle is the vector sum of the interfacial surface energies between the solid/liquid (γ_{sl}), liquid/vapor (γ_{lv}), and solid/vapor (γ_{sv}) phases. Young's equation represents a steady-state condition for a solid/liquid interface in stable or metastable thermodynamic equilibrium. Temperature changes have been shown to affect the contact angle of many different systems. The temperature effect, in most cases, can be explained by a reaction at the liquid/solid interface. Thermally activated reactions can occur because many systems are not at chemical equilibrium. The reactions that contribute to wetting (decrease of the contact angle) are those that increase the driving force for wetting ($\gamma_{sv} - \gamma_{SL}$), which is acting at the surface of the liquid drop and the solid substrate. The reactions that contribute to the driving force for wetting are the ones in which the composition of the substrate changes by dissolution of a component of the liquid. On the contrary, if the reaction results in a change of the liquid's composition by dissolution of the solid substrate, but with no change in the composition of the substrate, there is no contribution to the driving force for wetting.

As mentioned above, if the solid substrate is an active participant in the reaction, the free energy of the outer surface of the liquid drop will contribute to the driving force for wetting. As the drop expands on the substrate, the perimeter remains in contact with the unreacted solid and thus the reaction continues to contribute to the driving force for wetting. Examination of phase diagrams representing the interaction between the constituents of the liquid and solid surfaces can help to predict the wetting behavior of a system.

Moreover, measurement of wettability of powders consisting of irregular and polysized particles is extremely difficult. Several techniques have been proposed in the thermodynamic literature to measure wettability. However, these techniques have been applied mostly to non-metallic liquids and their application to metal ceramic systems with reference to pressure casting of composites has been quite limited. The engineering approaches to increasing wettability can be broadly classified into two categories. One method is the surface modification of the reinforcement phase and the other technique is melt treatment. Surface modifications of reinforcements include heat treatment of the particulates to determine surface gas desorbtion, surface oxidation and coating of particles with materials that react with the matrix. Melt treatment is usually done to promote reactivity between the metal and the particulate surface. The wetting reaction must be constrained to prevent reinforcement degradation during the fabrication of subsequent utilization (Ho and Wu, 1998).

2.8. Particulate reinforcement

The improvement in toughness due to the particulate reinforcement depends on the residual stresses surrounding the particles, the weight fraction of the particles, size and shape of the particles (Suery and Esperance, 1993). Particles can be spherical, disk-shaped, rod shaped, and plate shaped. Each particle forces the crack to go out of plane, and can force the crack to deflect in more than one direction and thus increase the fracture surface energy (Gogopsi, 1994). Plate and rod shaped particles can increase the composite toughness by another mechanism called as 'pullout' and 'bridging'. The residual stress around the particles results from thermal expansion mismatch between the particles and the matrix, which helps to resist the crack propagation. The term 'particulates' is used to distinguish these materials from particle and referred as a large, diverse group of materials that consist of minute particles. The second phase particle can produce small but significant increase in toughness and consequently increases its strength through crack deflection processes. The particles, sometimes given a proprietary coating can be used for improving strength. When compared to whiskers-reinforcement systems, particle reinforcement systems have less processing difficulties and should permit to add higher weight fractions of the reinforcing phase. The orientation of particles appears as flat plates (Matthew and Rawlings, 1999; Pardo et al., 2005).

3. Experimental procedure

3.1. Materials selected for processing composites

Aluminum – 11.8% silicon (LM6)

The main materials used in this project are LM6 aluminum alloy as a matrix material and SiO$_2$-quartz as a particulate reinforced added in different percentages. Pure (99.99%) aluminum has a specific gravity of 2.70 and its density is equals 2685kg/m^3. The details of the LM6 alloy properties and composition is shown in Table 1 and Table 2.

Composition LM6	
Al	85.95
Cu	0.2
Mg	0.1
Si	11.8
Fe	0.5
Mn	0.5
Ni	0.1
Zn	0.1
Lead	0.1
Tin	0.05
Titanium	0.2
Other	0.2

Table 1. Composition of LM6(Sayuti, Sulaiman, Baharudin, et al., 2011)

PHYSICAL PROPERTIES	VALUES
Density (g/cc)	2.66
MECHANICAL PROPERTIES	**VALUES**
Tensile strength, Ultimate (MPa)	290
Tensile Strength, Yield (MPa)	131
Elongation %; break (%)	3.5
Poisson's ratio	0.33
Fatigue Strength (MPa)	130
THERMAL PROPERTIES	**VALUES**
CTE, linear 20°C (μm/m-°C)	20.4
CTE, linear 250°C (μm/m-°C)	22.4
Heat Capacity (J/g- °C)	0.963
Thermal Conductivity (W/m-K)	155
Melting Point (°C)	574

Table 2. Physical, Mechanical and thermal properties of LM6 (Sulaiman, et al., 2008)

Quartz

Pure and fused silica is commonly called quartz. Quartz is a hard mineral which is abundantly available as a natural resource. It has a rhombohedra crystal structure with a hardness of 7 on the Mohs scale and has a low specific gravity ranging from 2.50 to 2.66. It provides excellent hardness when incorporated into the soft lead-alloy, thereby making it better suited for applications where hardness is desirable. It also imparts good corrosion resistance and high chemical stability. It is a mineral having a composition SiO_2, which is the most common among all the materials, and occurs in the combined and uncombined states. It is estimated that 60% of the earth's crust contain SiO_2. Sand, clays, and rocks are largely composed of small quartz crystals. SiO_2 is white in color in the purest form. The properties of pure quartz are listed in the Table 3.

Properties of quartz	
Molecular weight	60.08
Melting Point °C	1713
Boiling Point °C	2230
Density gm/cc	2.32
Thermal Conductivity	0.01 W/cm K (bulk)
Thermal Diffusivity	0.009 cm2/sec (bulk)
Mohs Hardness @ 20 °C	7 Modified Mohs
Si %	46.75
O %	53.25
Crystal Structure	Cubic
Mesh size	230
Size	65 microns (65 μm)

Table 3. Properties of quartz

Preparation of materials

The materials used in this work were Aluminum LM6 alloy as the matrix and SiO₂ as reinforcement particulates with different weight percentages. The tensile test specimens were prepared according to ASTM standards B 557 M-94 (ASTM, 1991). Sodium silicate and CO_2 gas was used to produce CO_2 sand mould for processing composite casting. The aluminum alloy, LM6, was based on British standards that conform to BS 1490-1988 LM6. Alloy of LM6 is actually a eutectic alloy having the lowest melting point that can be seen from the Al-Si phase diagram. The main composition of LM6 is about 85.95% of aluminum and 11.8% of silicon.

The SiO₂ particulate used as a second phase reinforcement in the alloy matrix was added on the molten LM6 by different weights fraction such as 5%, 10%, 15%, 20%, 25%, and 30%. The mesh size of Silicon Dioxide particulate is 230 microns and the average particle size equal to 65 microns (65μm).

Fabrication of composites

Only one type of pattern was used in this project and the procedure for making the pattern involves the preparation of drawing, selection of pattern material and surface finishing. Carbon dioxide moulding process was used to prepare the specimens as per the standard moulding procedure. Quartz-particulate reinforced MMCs were fabricated by casting technique. Six different weight fractions of SiO₂ particle in the range from 5%, 10%, 15%, 20%, 25%, and 30% by weight were used. In this research work, the particulates were preheated to 200 °C in a heat treatment muffle furnace for 2 hours and it was transferred immediately in the crucible containing liquid LM6 alloy.

3.2. Testing methods

Tensile testing

Tensile test was conducted to determine the mechanical properties of the processed SiO₂ particulate reinforced LM6 alloy composites. Test specimens were made in accordance to ASTM standard B557 M-94. A 250 KN servo hydraulic INSTRON 8500 UTM was used to conduct the tensile test. The tensile testing of the samples was performed based on the following specifications and procedures according to the ASTM standards, which of one crosshead speed of 2.00 mm/minute, grip distance 50.0 mm, specimen distance 50.0 mm and temperature 24 °C.

Hardness measurement

The hardness testing was done on a Rockwell Hardness Tester. The hardness of composites was tested by using MITUTOYO ATK-600 MODEL hardness tester. For each sample, ten hardness readings were taken randomly from the surface of the samples. Hardness values of different types of the processed composites are determined for different weight fraction % of titanium carbide particulate containing aluminum-11.8% silicon alloy and graphs were plotted between the hardness value and the corresponding type of particulate addition on weight fraction basis.

Impact testing

The impact test was conducted in accordance with ASTM E 23-05 standards at room temperature using izod impact tester. The casting processing steps and testing shows are shown in Figure 2.

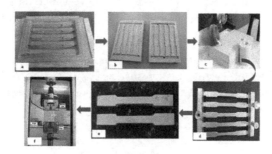

Figure 2. The casting processing steps; (a) Pattern of mould (b) sand mould : drag and copper (c) melting and pouring in the sand mould (d) tensile specimens with gating system (e) tensile specimen after removing of gating systems (f) tensile testing

Density measurement

The density of a material is defined as its mass per unit volume. A&D-GR 200 – Analytical Balance was used to conduct the density measurement. The theoretical density of each set of composites was calculated using the rule of mixtures (Rizkalla and Abdulwahed, 1996). Each pellet was weighed in air (W_a), then suspended in Xylene and weighed again (W). The density of the pellet was calculated according to the formula:

$$Density = \frac{Wa}{(Wa - Ww)} \times density\ of\ Xylene \tag{2}$$

Thermal diffusivity measurement

Thermal diffusivity of composite materials is measured using the photo flash method. The photoflash detection system consists of a light source, sample holder, thermocouple, low noise pre amplifier, oscilloscope, photodiode and a personal computer. The temperature rise at the back surface of the sample is detected by the thermocouple. The detected signal is amplified by a low-noise preamplifier and processed by a digital oscilloscope (Carter and Norton, 2007; Yu et al., 2002).

The voltage supplied to the camera flash is always maintained below 6 Volts before switching on the main power supply. The sample is machined to acquire flat surface to obtain better quality result and it is attached directly to the thermocouple. The camera flash is located at 2 cm in front of the sample holder. Before starting the equipment, the set up was tested using a standard material such as aluminium. Measurement was carried out every 10 minutes to allow the sample to thermally equilibrate at room temperature. The data was analyzed before running the next measurement.

Photoflash detection system is not an expensive method and the standard thermal diffusivity value for aluminum is equal to 0.83 cm^2/sec for thickness greater than 0.366 cm (Muta et al., 2003). In the photo flash system, the excitation source consists of a high intensity camera flash. This method is well suitable for aluminum, aluminum alloys and aluminum-silicon particulate metal matrix composites (Collieu and Powney, 1973). The thermal diffusivity values can be obtained for different thicknesses of the test samples. The thermal diffusivity α determines the speed of propagation of heat waves by conduction during changes of temperature with time. It can be related to α, the thermal conductivity through the following equation (Michot et al., 2008; Taylor, 1980).

The photo flash technique was originally described by Parker and it is one of the most common ways to measure the thermal diffusivity of the solid samples. The computer is programmed to calculate the thermal diffusivity, α, using the equation:

$$\alpha = \frac{\left(1.37 \, x \, L^2\right)}{[(3.14)^2 \, x t_{0.5}]} \tag{3}$$

Where L = thickness in mm and t $_{0.5}$ = half rise time in seconds.

Scanning Electron Microscopy (SEM)

LEO 1455 variable pressure scanning electron microscope with Inca 300 Energy Dispersive X-ray (EDX) was used to investigate the morphological features. Results and data obtained from the tensile tested samples were correlated with the reported mechanical properties for each volume fraction of silicon dioxide percentage addition to the LM6 alloy matrix.

4. Results

Tensile properties

The average value of tensile strength (MPa) and Young's Modulus (MPa) versus weight fraction of SiO$_2$ is shown in the Figures 3 and 4.

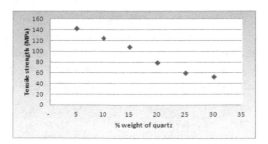

Figure 3. Tensile strength Vs % weight

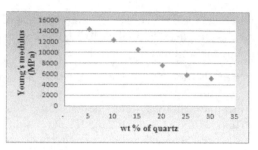

Figure 4. Young"s modulus Vs % weight

The graph plotted between the average tensile strength and modulus or elasticity values versus variation in weight fraction of quartz particulate addition to LM6 alloy indicates that both the properties decreases with increase in the addition of the quartz particulate. The increase of closed pores content with increasing quartz particulate content would create more sites for crack initiation and hence lower down the load bearing capacity of the composite. The fluctuation maybe due to the non-uniform distribution of quartz particulates, due to experimental errors and also depends on the cooling rate of the castings (ASTM, 1991; Seah, et al., 2003). When particulates increase, particles are no longer isolated by the ductile aluminum alloy matrix, therefore cracks will be not arrested by the ductile matrix and gaps would propagate easily between the quartz particulates. This residual stress affects the material properties around it and the crack tips and the fracture toughness values would be altered. Consequently, these residual stresses would probably contribute for the brittle nature of the composites. It should be noted that the compressive strength of the quartz particulate dominates which is more than the tensile strength of the LM6 alloy matrix and hence the tensile strength decreases with more amount of addition of quartz particulate afact which is well supported is well supported and evidenced from the literature citation (Rizkalla and Abdulwahed, 1996; Seah, et al., 2003).

Hardness

Similarly, for a given SiO_2 reinforcement content, some differences in the hardness values were observed depending upon the particle size of the constituents. From the Table 4, data on hardness of quartz particulate reinforced composites made in sand mold is listed. It was found that the hardness value increased gradually with the increased addition of quartz particulate by weight fraction percentage as shown in Figure 5.

The maximum hardness value obtained based on the Rockwell superficial 15N-S scale was 67.85 for 30% weight fraction addition. The EDS spectrums for 30% wt of SiO_2 are shown in Figure 5. Their respective elemental analysis is shown in Table 4. It was observed that the grain-refined composite casting has higher weight percentage of Si compared with the original LM6 casting. These results indicate the interrelationship between the thermal properties and hardness.

Impact strength

Impact strength data of quartz particulate reinforced composite castings processed was determined and it is listed in the Table 4. From the plotted graph shown in the Figure 6, it is found that the impact strength values were gradually increased with the increased addition of quartz particulate in the alloy matrix. The maximum value of impact strength was 24.80 N-m for 30% weight fraction addition of quartz particulate to the alloy matrix. A reason for the increased volume impact–abrasive wear of the SiO2 particle reinforced composites lies in the propensity of the carbides to fracture and spall as a result of the repeated impact from the quartzite. In the monolithic ferrous-based alloys, the matrix can absorb substantial damage in the form of plastic deformation. This plastic deformation is in fact beneficial in that, the matrix will get harder as a result, and wear, fatigue type processes ending as a material removal mechanism. In the SiO₂ particle reinforced composites, however, the high weight fraction of SiO₂ limits the amount of plastic deformation that the matrix can absorb. This leads more quickly to SiO₂ reinforcement fracture, matrix– SiO₂ particle delamination, and S₁O₂ particle spalling. As a consequence, volume impact–abrasive wear increases at a more rapid rate for the composite materials as the hardness increases. However, for the very 'hardest' S₁O₂ particle reinforced composites, impact–abrasion resistance is very good. The summary of mechanical properties of quartz particulate reinforced composite castings processed was determined and it is listed in the Table 4.

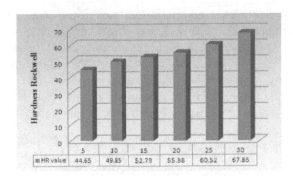

Figure 5. Hardness Vs wt % of quartz

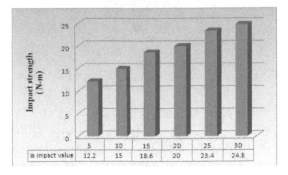

Figure 6. Impact strength Vs Weight fraction % of quartz

Wt % of quartz	UTS (MPa)	Yield (MPa)	Young's modulus MPa	Fracture stress MPa	Ductility %	Reduction in area %	Rockwell Hardness	Impact (N-m)
5 %	142.99	132.00	14351	189.50	1.214	2.863	44.65	12.20
10 %	124.74	129.60	12350	164.60	1.412	2.864	49.85	15.00
15 %	108.47	118.50	10635	142.20	1.422	3.042	52.73	18.60
20 %	78.97	109.60	7621	128.40	1.632	3.264	55.38	20.00
25%	59.53	100.50	5853	115.30	1.824	3.625	60.52	23.40
30%	52.64	92.65	5242	104.60	1.741	3.482	67.85	24.80

Table 4. Mechanical properties of quartz particulate composites

Density

Figure 8 gives the influence of quartz addition on the density. The graph shows that as the quartz-silicon dioxide content was gradually increased, the density of the Aluminum composite decreased. Slight decrease was observed in the density because quartz-silicon dioxide has a slight lower density value than LM6 (the density of LM6 is 2.65grs/cc and of quartz is 2.23grs/cc).

The investigation of the aluminum composite was well documented. The percentage of the closed pores in the sintered composites increased with increasing quartz content. This can be attributed to silica being harder than aluminum and non deformation at all under the applied compaction load. The morphological features of quartz particles were significantly different from those of Aluminum and as a result, the interparticle friction effects were different. Therefore, the increase in the amount of closed pores with increasing quartz content would justify the observed decrease in density (Rizkalla and Abdulwahed, 1996).

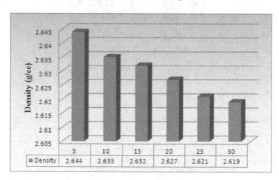

Figure 7. Graph plotted on density versus %wt fraction of SiO_2

Thermal properties

Quartz particulate reinforced composite castings made in grey cast iron mold were tested and analyzed for thermal properties. Graphs are plotted between the weight fraction % addition of quartz and thermal diffusivity and thermal conductivity values. It is found that the thermal diffusivity of the quartz composites decreased with the increased addition in the alloy matrix. Reversely, the thermal conductivity of the quartz composites decreased with the increased addition of quartz particulate in the alloy matrix. Quartz particulates are a ceramic reinforcement phase and on addition of this in the alloy matrix reduces the thermal conductivity. The data for thermal diffusivity and thermal conductivity of the quartz particulate reinforced composites made in sand mold is given in the Table 4. These are illustrated in the plotted graphs and are shown in Figure 8 and 9. The thermal diffusivity and thermal conductivity for 30% weight fraction addition of quartz are 0.2306 cm²/sec and 52.9543 W/mK respectively and it is well supported from the literature citation (Collieu and Powney, 1973). The summary of physical properties of quartz particulate reinforced composite castings processed was determined and it is listed in the Table 5.

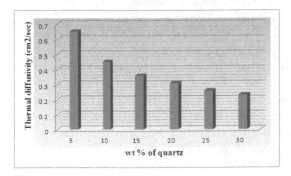

Figure 8. Thermal diffusivity Vs Wt Fraction % of quartz

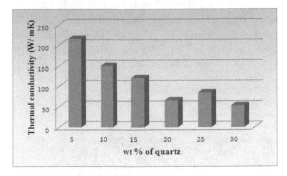

Figure 9. Thermal conductivity Vs Wt Fraction % of quartz

Wt % of quartz	Density (g/cc)	Thermal diffusivity cm²/sec	Thermal conductivity (W/ mK)
5 %	2.644	0.6513	215.826
10 %	2.635	0.4514	149.584
15 %	2.632	0.3595	119.933
20 %	2.627	0.3102	65.6860
25%	2.621	0.2590	84.6830
30%	2.619	0.2306	52.9543

Table 5. Physical properties of quartz particulate composites

Scanning electron microscopy (SEM)

Scanning Electron Microscopy and energy dispersive spectroscopy was employed to obtain some qualitative evidences on the particle distribution in the matrix and bonding quality between the particulate and the matrix. Besides this the fracture surface of the composite was analyzed by using SEM to show the detail of chemically reacted interfaces. Thus, in order to increase the potential application of MMCs, it is necessary to concentrate on the major aspects, like particle size of quartz and quartz distribution concentration.

The fracture surfaces or fractographs are shown in the Figures 10-15 after tensile testing the specimens having different weight fraction of quartz particulate. It was observed that the increase of SiO_2 content would create more sites for crack initiation and would lower the load bearing capacity of MMCs. In addition the number of contacts between quartz particles would increase and more particles were no longer isolated by the ductile aluminum alloy matrix. Therefore, cracks were not arrested by the ductile matrix and they would propagate easily between quartz particulates. Decrease of SiO_2 content to less than 30% in the matrix and a particle size of 230 micron could increase the tensile strength. Hence cracking on the surface is not too dominant. This phenomenon is shown in Figure 10. The problem on interfacial bonding between the particulate quartz and the matrix during the solidification of composites can be ignored because the phenomena of cracking occurs only in a small part of the surface (Seah, et al., 2003). In contrast, when the content of quartz was increased (30%), interfacial bonding concept would be an important phenomenon because the surface cracking will be distributed on the surface of the parts. The other problem caused by the interaction between Aluminum alloy and quartz particle is not a significant one and it is removed while solidification during the pouring process and due to slip inter bonding/ inter granular movement which is illustrated with the aid of Figure 11.

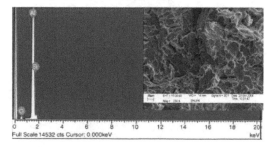

Figure 10. EDX Spectrum and Fractograph of 5wt% quartz particulate reinforced in quartz -LM6 alloy matrix composite at 250X magnification by SEM after tensile testing.

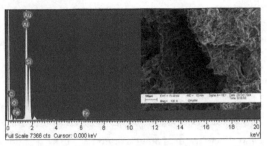

Figure 11. EDX Spectrum and Fractograph of 10wt% quartz particulate reinforced in quartz -LM6 alloy matrix composite at 100X magnification by SEM after tensile testing.

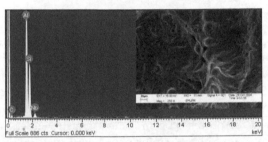

Figure 12. EDX Spectrum and Fractograph of 15wt% quartz particulate reinforced in quartz -LM6 alloy matrix composite at 250X magnification by SEM after tensile testing.

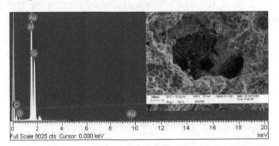

Figure 13. EDX Spectrum and Fractograph of 20wt% quartz particulate reinforced in quartz -LM6 alloy matrix composite at 100X magnification by SEM after tensile testing

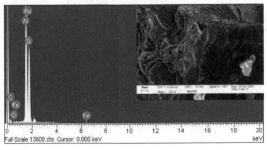

Figure 14. EDX Spectrum and Fractograph of 25wt% quartz particulate reinforced in quartz -LM6 alloy matrix composite at 250X magnification by SEM after tensile testing

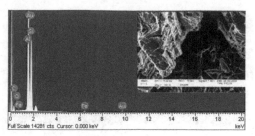

Figure 15. EDX Spectrum and Fractograph of 30wt% quartz articulate reinforced in quartz -LM6 alloy matrix composite at 250X magnification by SEM after tensile testing

5. Conclusions

In this study, the compressive strength of the silicon dioxide particulate reinforcement dominates and influences more effectively than the tensile strength of the LM6 alloy matrix phase. Hence the values of tensile strength and modulus of elasticity are decreased with the increased addition of SiO_2 particulate from 5 to 30% by volume fraction basis. This fact from the present experimental research is well supported and validated from the literature. The mechanical behaviour of the processed composite had a strong dependence on the volume fraction addition of the second phase reinforcement particulate on the alloy matrix. On the other hand, decreasing the SiO_2 particulate content less than 30% by weight along with the particle size constraint as 230 mesh-65 microns would increase the tensile strength and cracking on the surface might not be too dominant. The hardness value of the silicon reinforced aluminum silicon alloy matrix composite is increased with the addition of quartz particulate in the matrix.

The density of these composites decreased slightly with increasing quartz content. Slight decrease was observed in the density because quartz-silicon dioxide has a slightly lower density value than LM6. For a given particle size combination, the thermal diffusivity and thermal conductivity decreases as SiO_2 wt % of the composite increases. The particle size ratio of the constituents becomes an important factor for thermal properties, especially above 10wt. % SiO_2. A higher Al/ SiO_2 particle size ratio results in segregation of SiO_2 particles along the LM6 boundaries. This yields lower thermal conductivity with respect to the homogeneously distributed reinforcement. Therefore, a thermal conductivity value that is less than the expected one might be attributed to the micro-porosity in the segregated structure. Similar tendencies were also observed for the results of hardness tests.

In future, it is strongly recommended that tensile tests be performed by reinforcing the second phase quartz particulate addition to the LM6 alloy matrix by limiting it up to 15 wt%. In addition, compressive strengths testing of the processed composite samples can be done to highlight the benefits, advantages and applications of these composites. It is also worthwhile to conduct heat treatment studies of these processed composites and this will be in the scope of future research work.

Author details

M. Sayuti
Department of Mechanical and Manufacturing Engineering, Faculty of Engineering, Universiti Putra Malaysia, Serdang, Selangor, Malaysia
Department of Industrial Engineering, Faculty of Engineering, Malikussaleh University, Lhokseumawe, Aceh - Indonesia

S. Sulaiman, B.T.H.T. Baharudin and M.K.A. Arifin
Department of Mechanical and Manufacturing Engineering, Faculty of Engineering, Universiti Putra Malaysia, Serdang, Selangor, Malaysia

T.R. Vijayaram
Faculty of Engineering and Technology (FET) Multimedia University, Jalan Ayer Keroh Lama, Bukit Beruang, Melaka. Malaysia

Acknowledgement

The authors would like to express their deep gratitude and sincere thanks to the Department of Mechanical and Manufacturing Engineering, Universiti Putra Malaysia for their help to complete this work.

6. References

Adams, D. F., carlsson, L. A., and Pipes, R. B. (2003). *Expeirmental Characterization of Advanced Composite Materials* (3rd ed.). Florida, USA: CRC Press LLC.

Ashby, M. F., and Jones, D. R. H. (1980). *Engineering Materials: An Introduction to their properties and applications, International series on materials science and technology.* UK: Elsevier Science and Technology.

Asthana, R., and Rohatgi, P. K. (1993). A Study of Metal-Ceramic Wettability in Sic-Al Using Dynamic Melt Infiltration of SiC. *Key Engineering Materials, 70-80,* 47-62.

ASTM. (1991). American Society for Testing and Material, *Anual Books of ASTM Standards.* USA.

Baker, A. R., Dawson, D. J., and Evans, D. C. (1987). Ceramics and Composite, Materials for Precision Engine Components. *Journal of Materials & Design, 8*(6), 315-323.

Basavarajappa, S., Chandramohan, G., and Dinesh, A. (2004). *Mechanical Properties of MMC'S-an Experimental Inversitgation.* Paper presented at the International Symposium of research Students on Materials and Engineering, Chennai.

Burr, A., Yang, J. Y., Levi, C. G., and Leckie, F. A. (1995). The strength of metal-matrix composite joints. *Acta Metallurgica et Materialia, 43*(9), 3361-3373.

Carter, C. B., and Norton, M. G. (2007). *Ceramic Material: Science and Engineering* New York: Springer.

Chambers, B. V., Seleznev, M. L., A.Cornie, J., Zhang, S., and Rye, M. A. (1996). The Strength and Toughness of Cast Aluminium Composite as a Function of Composition, Heat Treatment and Particulate. *SAE International Journal*, 164-169.

Clyne, T. W. (1996). Interfacial Effects in Particulate, Fibrous and Layered Composite Materials. *Key Engineering Materials, 116-117*, 133-152.

Collieu, A. M. B., and Powney, D. J. (1973). *The Mechanical and Thermal Properties of Materials* UK, London: Edward Arnold (Publishers) Ltd.

Daniel, I. M., and Ishai, O. (1994). *Engineering Mechanics of Composite Materials*. USA: Oxford University Press.

Dusza, J., and Sajgalik, P. (1995). Fracture Toughness and Strength Testing of Ceramic Composites. In N. P, Cheremisinoff and P. N.Cheremisinoff (Eds.), *Handbook of Advanced Materials Testing* (pp. 399-435). New York, USA: Marcel Dekker, Inc.

El-Mahallawy, N. A., and Taha, M. A. (1993). Reinforcement Considerations for High Temperature Metal Matrix Composites. *Key Engineering Materials, 79-80*, 1-14.

Feest, E. A. (1986). Metal matrix composites for industrial application. *Materials & Design, 7*(2), 58-64.

Fridlyander, J. N. (1995). *Metal Matrix Composites*. UK, London: Chapman & Hall.

Gay, D., V.Hoa, S., and W.Tsai, S. (2003). *Composite Materials: Design and Applications*. USA: CRC Press LLC.

Gogopsi, Y. G. (1994). Particulate Silicon Nitride - Based Composite. *Materials science, 29*(4), 2541-2556.

Gregolin, Goldenstein, H., Gonçalves, M. d. C., and Santos., R. G. d. (2002). Aluminium Matrix Composites Reinforced with Co-continuous Interfaced Phases Aluminium-alumina Needles. *Materials Research, 5*(3), 337-342.

Hashim, J., Looney, L., and Hashmi, M. S. J. (1999). Metal matrix composites: production by the stir casting method. *Journal of Materials Processing Technology, 92-93*, 1-7.

Hashim, J., Looney, L., and Hashmi, M. S. J. (2001). The enhancement of wettability of SiC particles in cast aluminium matrix composites. *Journal of Materials Processing Technology, 119*(1-3), 329-335.

Hashim, J., Looney, L., and Hashmi, M. S. J. (2002). Particle distribution in cast metal matrix composites--Part I. *Journal of Materials Processing Technology, 123*(2), 251-257.

Ho, H.-N., and Wu, S.-T. (1998). The wettability of molten aluminum on sintered aluminum nitride substrate. *Materials Science and Engineering: A, 248*(1-2), 120-124.

Huda, D., El Baradie, M. A., and Hashmi, M. S. J. (1993). Metal-matrix composites: Manufacturing aspects. Part I. *Journal of Materials Processing Technology, 37*(1-4), 513-528.

Humphreys, J. (1987). Composites for automotive on-engine applications. *Materials & Design, 8*(3), 147-151.

Kaczmar, J. W., Pietrzak, K., and Wlosinski, W. (2000). The production and application of metal matrix composite materials. *Journal of Materials Processing Technology, 106*(1-3), 58-67.

Kok, M. (2005). Production and mechanical properties of Al2O3 particle-reinforced 2024 aluminium alloy composites. *Journal of Materials Processing Technology, 161*(3), 381-387.

Matthew, F. L., and Rawlings, R. D. (1999). *Composite Material; Engineering and Science.* UK: Imperial College of Science.

Mazumdar, S. K. (2002). *Composites Manufacturing: Materials, Product and Process Engineering.* USA: CRC Press Inc.

Michot, A. e., Smith, D. S., Degot, S., and Gault, C. (2008). Thermal Conductivity and Specific Heat of Kaolinite: Evolution with Thermal Treatment. *Journal of the European Ceramic Society 28,* 2639-2644.

Mortensen, A., Cornie, J. A., and Flemings., M. C. (1989). Solidification Processing of Metal-Matrix Composites. *Materials & Design, 10*(2), 68-76.

Muta, H., Kurosaki, K., Uno, M., and Yamanaka, S. (2003). Thermoelectric properties of constantan/spherical SiO2 and Al2O3 particles composite. *Journal of Alloys and Compounds, 359*(1-2), 326-329.

Oh, S.-Y., Cornie, J. A., and Russell., K. C. (1987). *Particulate Wetting and Metal: Ceramic Interface Phenomena.* Paper presented at the Ceramic Engineering Science Proceedings.

Pardo, A., Merino, M. C., Merino, S., Viejo, F., Carboneras, M., and Arrabal, R. (2005). Influence of reinforcement proportion and matrix composition on pitting corrosion behaviour of cast aluminium matrix composites (A3xx.x/SiCp). *Corrosion Science, 47*(7), 1750-1764.

Prasad, S. V., and Asthana, R. (2004). Aluminium Metal Matrix Composite for Automotive Applications: Tribological Considerations. *Tribology Letters, 17*(3), 445-453.

Rajan, T. P. D., Pillai, R. M., and Pai, B. C. (1998). Reinforcement coatings and interfaces in aluminium metal matrix composites. *Journal of Materials Science, 33*(14), 3491-3503.

Rizkalla, H. L., and Abdulwahed, A. (1996). Some mechanical properties of metal-nonmetal Al---SiO2 particulate composites. *Journal of Materials Processing Technology, 56*(1-4), 398-403.

Rohatgi, P. K. (1993). Metal Matrix Composite, Casting Processes. *Science Journal, 43*(4), 323-349.

Sahin, Y. (2003). Preparation and some properties of SiC particle reinforced aluminium alloy composites. *Materials & Design, 24*(8), 671-679.

Sayuti, M., Sulaiman, S., Baharudin, B. T. H. T., Arifin, M. K. A., Suraya, S., and Vijayaram, T. R. (2010). *Mechanical properties of particulate reinforced aluminium alloy matrix composite.*

Sayuti, M., Sulaiman, S., Baharudin, B. T. H. T., Arifin, M. K. A., Vijayaram, T. R., and Suraya, S. (2011) Influence of mechanical vibration moulding process on the tensile properties of TiC reinforced LM6 alloy composite castings. *Vol. 66-68* (pp. 1207-1212).

Sayuti, M., Sulaiman, S., Vijayaram, T. R., Baharudin, B. T. H. T., and Arifin, M. K. A. (2011) The influence of mechanical vibration moulding process on thermal conductivity and diffusivity of Al-TiC particulate reinforced composites. *Vol. 311-313* (pp. 3-8).

Sayuti, M., Suraya, S., Sulaiman, S., Vijayaram, T. R., Arifin, M. K. H., and Baharudin, B. T. H. T. (2011) Thermal investigation of aluminium - 11.8% silicon (LM6) reinforced SiO2 - Particles. *Vol. 264-265* (pp. 620-625).

Schey, J. A. (2000). *Introduction to Manufacturing Processes* (Vol. 3). USA: McGraw Hill.

Seah, K. H. W., Hemanth, J., and Sharma, S. C. (2003). Mechanical properties of aluminum/quartz particulate composites cast using metallic and non-metallic chills. *Materials & Design,* 24(2), 87-93.

Sharma, S., Seah, K. H. W., Girish, B. M., Kamath, R., and Satish, B. M. (1997). Mechanical properties and fractography of cast lead-alloy/quartz particulate composites. *Materials & Design, 18*(3), 149-153.

Singh, M., Mondal, D. P., Jha, A. K., Das, S., and Yegneswaran, A. H. (2001). Preparation and properties of cast aluminium alloy-sillimanite particle composite. *Composites Part A: Applied Science and Manufacturing, 32*(6), 787-795.

Suery, M., and Esperance, G. L. (1993). Interfacial Reactions and Mechanical Behaviour of Aluminium Matrix Composites Reinforced with Ceramic Particles. *Key Engineering Materials, 79-80,* 33-46.

Sulaiman, S., Sayuti, M., and Samin, R. (2008). Mechanical properties of the as-cast quartz particulate reinforced LM6 alloy matrix composites. *Journal of Materials Processing Technology, 201*(1-3), 731-735.

Surappa, M. (2003). Aluminium matrix composites: Challenges and opportunities. *Sadhana, 28*(1), 319-334.

Taylor, R. (1980). Construction of apparatus for heat pulse thermal diffusivity measurements from 300-3000K *Journal of physics E : Scientific Instruments, 13*(11).

Therén, K., and Lundin, A. (1990). Advanced composite materials for road vehicles. *Materials & Design, 11*(2), 71-75.

Upadhyaya, G. S. (1990). Trends in advanced materials and processes. *Materials & Design, 11*(4), 171-179.

Vijayaram, T. R., Sulaiman, S., Hamouda, A. M. S., and Ahmad, M. H. M. (2006). Fabrication of fiber reinforced metal matrix composites by squeeze casting technology. *Journal of Materials Processing Technology, 178*(1-3), 34-38.

Wang, J.-j., Guo, J.-h., and Chen, L.-q. (2006). TiC/AZ91D composites fabricated by in situ reactive infiltration process and its tensile deformation. *Transactions of Nonferrous Metals Society of China, 16*(4), 892-896.

Weeton, J. W., Peters, D. M., and Thomas, K. L. (1988) Engineers' Guide to Composite Materials. Metals Park, Ohio 44073, USA: American Society for Metals.

Yu, S., Hing, P., and Hu, X. (2002). Thermal conductivity of polystyrene-aluminum nitride composite. *Composites Part A: Applied Science and Manufacturing, 33*(2), 289-292.

Zhu, S. J., and Iizuka, T. (2003). Fabrication and mechanical behavior of Al matrix composites reinforced with porous ceramic of in situ grown whisker framework. *Materials Science and Engineering A, 354*(1-2), 306-314.

Chapter 8

YSZ Reinforced Ni-P Composite by Electroless Nickel Co-Deposition

Nor Bahiyah Baba

Additional information is available at the end of the chapter

1. Introduction

The importance of nickel ceramic composite in engineering applications especially for corrosion (Rabizadeh and Allahkaram 2011), wear (Lekka, Zanella et al. 2010) and thermal (Gengler, Muratore et al. 2010) resistance, and also for fuel cell anode (Pratihar, Sharma et al. 2007) has increasingly significance. Nickel ceramic composite namely Ni-P-YSZ manufactured by electroless nickel (EN) co-deposition process is investigated for its composite ratio, porosity content and electrical conductivity. The composite is the state-of-the-art as it is the combination of nickel that is very well known for its high thermal, electronic conductivity and corrosion resistance with yttria-stabilised zirconia (YSZ) for its high ionic conductivity, impact and hardness strength. The understanding of this composite formulation and properties will increase its ability as well as expanding its application in multi-engineering disciplines.

The unique of this composite is the fabrication method via EN co-deposition process. This is a single fabrication method consists of an in-situ incorporation of inert ceramic particles in the conventional Ni-P matrix. The co-deposition of fine particle in-situ in an electroless metal-matrix is very attractive as it saves energy and time. Typical co-deposition consists of particulates in the size range of 0.1-10 μm with loading of up to 40 vol.% of the total matrix (Feldstein 1990). Common fabrications for particulate composites are limited to conventional ceramic processing, solid state powder processing and thermal spraying techniques. The incorporation of particles in EN deposit has been widely investigated and the application of ceramic YSZ in EN composite coating is the new approach (Baba, Waugh et al. 2009).

EN deposition is an autocatalytic electrochemical reaction by a chemical reduction of nickel ions near the surface of the activated substrate (Feldstein 1990). In addition to its good coating characteristics, this process can be applied to the surfaces of almost all materials. The understanding of EN chemical process as described by equations (1) - (4) is the key to control the EN deposition morphology, composition and properties. The conventional EN deposition processes are a direct chemical reduction of nickel ions from EN Slotonip 1850 (Schloetter 2006) solution to metallic nickel process as illustrate in equation (2) and the deposition of phosphorus in equation (3) below. This process is controlled by the EN bath composition, bath temperature, bath pH and soaking time, deposition rate, substrate surface and substrate orientation. Altering the process parameters cause physical changes to the deposit.

$$(H_2PO_2)^- + H_2O \rightarrow H^+ + (HPO_3)^{2-} + 2H_{abs} \cdots \tag{1}$$

$$Ni^{2+} + 2H_{abs} \rightarrow Ni + 2H^+ \cdots \tag{2}$$

$$(H_2PO_2)^- + H_{abs} \rightarrow H_2O + OH^- + P \cdots \tag{3}$$

$$(H_2PO_2)^- + H_2O \rightarrow H^+ + (HPO_3)^{2-} + H_2 \cdots \tag{4}$$

The application of EN deposition in producing composites by in-situ incorporation of inert particles in the conventional Ni-P matrix has been extensively researched, i.e. incorporation of diamond, silicon carbide, silicon nitride, silicon oxide, boron carbide, alumina, iron oxide, titanium oxide, ceria, yttria, zirconia and PTFE particles with varying the particulate sizes (micro or nano), as listed in Table 1.

The composition of nickel and YSZ must be controlled to result in the desired properties. High ceramic composition ensures high wear, thermal and corrosion resistance. At the same time, the coefficient of thermal expansion within the coating and between coating and substrate should be compatible to avoid cracking and delamination respectively. Finally, the proportion of ceramic YSZ coating in the composite can be varied from layer to layer. The construction of a gradient of coating layers, for example, with lower ceramic content inside, at the lowest layer, increasing through subsequent layers confers the advantages of heat and corrosion resistance.

In order to effectively control the composition of Ni-YSZ composites for the best results, the conventional EN deposition process must be understood in terms of the process parameters especially the EN bath chemical and substrate conditions that affect its efficiency and quality. In addition to bath composition, conventional EN deposition is also controlled by several other bath-related factors such as bath temperature, bath pH and soaking time (Baudrand 1994).

The properties and affecting factors of conventional EN deposition might or might not be applicable to EN co-deposition. A study showed that the structural characteristics and phase transformation of EN co-deposition incorporating Si_3N_4, CeO_2 and TiO_2 remained unchanged as from those of the conventional EN deposition. In general, various factors have been shown to affect the deposition of EN composites, including (1) particle catalytic inertness, (2) particle charge, (3) EN bath composition, (4) bath reactivity, (5) particle compatibility with the matrix, (6) plating rate, and (7) particle size distribution (Feldstein 1990).

Types	Particle	Molecular formula	References
Inert/ hard materials	Diamond	C	(Hung, Lin et al. 2008); (Matsubara, Abe et al. 2007); (Sheela and Pushpavanam 2002)
	Silicon nitride	Si_3N_4	(Balaraju and Rajam 2008); (Dai, Liu et al. 2009); (Das, Limaye et al. 2007)
	Silicon carbide	SiC	(Berkh, Eskin et al. March 1996); (Kalantary, Holbrook et al. 1993); (Lin, Chen et al. 2006)
	Silicon oxide	SiO	(Dong, Chen et al. 2009)
	Boron carbide	B_4C	(Vaghefi, Saatchi et al. 2003)
	Alumina	Al_2O_3	(Balaraju, Kalavati et al. 2006); (Hazan, Reutera et al. 2008; Hazan, Werner et al. 2008)
	Ceria	CeO_2	(Necula, Apachitei et al. 2007)
	Yttria	Y_2O_3	(McCormack, Pomeroy et al. 2003)
	Zirconia	ZrO_2	(Shibli, Dilimon et al. 2006)
	Iron oxide	Fe_3O_4	(Zuleta, Galvis et al. 2009)
Others	Polytetrafluoroethylene	PTFE	(Ger and Hwang 2002)
	Titanium oxide	TiO_2	(Balaraju, Narayanan et al. 2006)
Multiple	Silicon carbide- Alumina	$SiC-Al_2O_3$	(Li, An et al. 2005; Li, An et al. 2005; Li, An et al. 2006)
	Silicon carbide-Graphite	SiC-G	(Wu, Shen et al. 2006)
	Silicon carbide- PTFE	SiC-PTFE	(Huang, Zeng et al. 2003)
	Zirconia-Alumina- Zirconium aluminide	$ZrO_2-Al_2O_3-Al_3Zr$	(Sharma, Agarwala et al. 2005)

Table 1. Types, names and formulas of the particles used in EN co-deposition

Particle stability in this case could be the charge stability of the particles in the solutions. Particle stability determines the particle dispersion in the solution and particles' tendency for agglomeration or sendimentation. Necula et al. (2007) found that particles are having good dispersion stability in deionised water but not so in the EN solution. Studies on alumina (Hazan et al. 2008b, 2008a), boron carbide (Vaghefi et al. 2003) and ceria (Necula et al. 2007) particles showed that the dispersion stability strongly depends on pH and the low stability caused a short sedimentation time. Studies by Hazan et al. (2008b; 2008a) on dispersion stability in Ni-P-Al_2O_3 EN system incorporating comb-polyelectrolyte showed high particle concentrations of up to 50 vol.% particle incorporation.

Periene et al. (1994) concluded that volume percent of co-deposition particles is dependent on powder conductivities and hydrophobic/ hydrophilic properties. Another study done on co-depositing boron carbide (B_4C) with particle sizes ranging from 5 to 11 µm in hypophosphite-reduced EN solution gave a maximum of 33 vol.% B_4C when the B_4C particles were wetted with surfactant before being added into the bath (Vaghefi, Saatchi et al. 2003). The surfactant is a blend of surface active agent which contains both hydrophilic and hydrophobic groups which helps increase deposition, even at 8 gl^{-1} particle loading. The application of surfactant in the deposition of PTFE on low carbon steel substrate showed strong adsorption (Ger and Hwang 2002).

The shape and size of the particles play an important role as they influence the deposition surface area and energy. It was found that spherical particles with smaller particle size (average 1 µm boron particles and 3.4 µm alumina particles) gave high particle concentration in the matrix (Apachitei, Duszczyk et al. 1998). Study done Balaraju et al. (2006a) varying alumina powder sizes of 50 nm, 0.3 µm and 1.0 µm showed that the highest particle incorporation occurred at 1.0 µm particle size.

Another important factor is the particle loading. Particle loading is the amount of powder particles in a litre solution. Co-depositing very fine polycrystalline diamond ranging between 8 and 12 µm with varying concentrations from 2-10 g/l onto aluminium substrate at 70-90°C for an hour yielded as high as 18.40 vol.% diamond powder in the deposit (Sheela and Pushpavanam 2002). The particle incorporation in a Ni-P-ZrO_2 EN system was found to be directly proportional to increase particle loading up to 9 g/l as well as the deposition rate (Shibli, Dilimon et al. 2006). In Ni-P-B_4C EN system, particle composition in the matrix increased from 12 to 33 vol.% as the particle loading increased from 1 to 8 g/l (Vaghefi, Saatchi et al. 2003). It was found that a particle loading more of than 15 g/l SiC in hypophosphite-reduced solution caused extensive bath foaming which reduced the plating rate (Kalantary, Holbrook et al. 1993).

Bath conditions might be disturbed by the addition of inert particles. The addition of silicon carbide (SiC) powders with particle size ranging from 4-7 µm in varying EN solution conditions (composition, pH, temperature and time) showed an increase in SiC loading in the deposition but a reduction of the deposition rate and deposition weight (Kalantary, Holbrook et al. 1993). Another study showed that co-deposition of SiC (1-5 µm) in sodium hypophosphite-reduced EN solution at pH 4.5-5.5, 80-90°C with air agitation resulted in 25-30 vol.% SiC in the deposition (Li 1997). Aggressive agitation might cause substrate or deposition abrasion by other factors such as particle hardness, particle shape and size, particle loading and bath movement (Kalantary, Holbrook et al. 1993).

Substrate orientation is defined as the position of the substrate in the EN bath during the EN deposition. The variations of substrate position have shown to give an effect on the EN deposition. In a study by Sheela & Pushpavanam (2002) on diamond EN co-deposition in hypophosphite-reduced solution showed that a vertical substrate position gave less than 20% particle incorporation compared to the horizontal position. Another study on Ni-P-SiC using hypophosphite-reduced solution with a particle loading of 25 g/l showed that the substrate held tangentially gave the highest particle composition in the matrix and a vertical position leading to uniform particle incorporation and adherence provided, that uniform agitation was used (Kalantary et al. 1993).

Having discussed all the above affecting factors on EN co-deposition process, the study is concentrated on four of the factors, namely, particle size (Balaraju, Kalavati et al. 2006), agitation method (Sevugan, Selvam et al. 1993), bath pH (Liu, Hsieh et al. 2006), and substrate surface treatment (Teixeira and Santini 2005). The application of design of experiment (DoE) using full factorials of a two-level factor is economical and practical to reduce the number of experiments. Unlike the Taguchi method, full factorial DoE highlights the significant main and interactions effects between the factors (Anthony 2003).

Factorials design is the basis of DoE. Factorials design is the most efficient way to study the effects of two or more factors. For example, 2-levels is simplified as 2^k or three levels as 3^k. Factorial design investigates all possible combinations of factor levels for each complete trial or replication of experiment. This is because the factors arranged in a factorial design are crossed by the Yates algorithm (Bisgaard 1998). A 2^k factorial design is used to study the effects of k factors with two levels for each factor. The k represents a multiple-factor design with a variation of treatment designs where a set of treatments (*factors*) are tested over one or more sets of treatments (*levels*). In practice the higher-order interactions are usually not significant, thus most design are limited to 2-3 levels.

The effects of EN composite coating parameters on Ni-YSZ composition and porosity content were investigated and optimized using full factorials and ANOVA. An empirical model for prediction of Ni composition and porosity content were established by means of piecewise linear regression analysis. The composite response to electrical conductivity is also investigated.

2. Experimental methods

2.1. Materials and preparations

A ceramic substrate of alumina tile (Coors Ceramics, U.K.) with manufacturer standard dimensions of 50 x 50 x 1 mm as shown in Figure 1 was used as the base for composite deposition. Surfaces of the substrate sample were treated by chemical etching and mechanical blasting. Chemical etching of the substrate sample was done by immersion for 5 minutes at room temperature in a hydrofluoric (HF) etching solution comprising a mixture of 1 part hydrofluoric acid (20ml/l) to 5 parts ammonium fluoride (NH4F) (2g/l). Then mechanical treatment of the substrate was done by blasting the substrate with brown alumina sand for 1-2 minutes. The contamination after sand blasting was cleaned ultrasonically for 30 minutes at room temperature by submerging the substrate samples individually in beakers filled with acetone.

Reinforcement ceramic particles of 8 mol% YSZ (8YSZ; United Ceramics, England) were used. Yttria stabilised zirconia (YSZ) is a ceramic phase that is very well-known for high hardness, good scratch and corrosion resistance; and very high thermal resistance, but low toughness and high brittleness. Normally the addition of ceramic particles in composites increases the composite's mechanical properties in term of its hardness, corrosion resistance and thermal resistance. The 8YSZ particles varied between the nominal sizes of 2 µm and 10 µm.

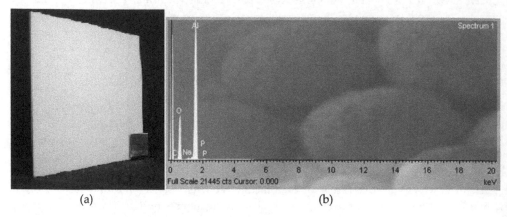

(a)   (b)

Figure 1. (a) Alumina tile as a ceramic substrate for EN co-deposition (b) Alumina substrate EDX spectrum

2.2. EN co-deposition

The ceramic alumina substrate requires sensitising to activate the surface. All non-proprietary solutions were prepared using AR grade chemicals and high purity deionised water. After the pre-treatment process sequence as listed in Table 2, the EN composite deposition of Ni-YSZ was performed within 3 hours to minimize effects of chemical degradation. The EN chemicals produced a bright mid-phosphorous (6 – 9%) nickel deposit. The solution was heated and the temperature maintained at 89°C using a Jenway hotplate.

Trade name	Soaking Time	Temperature
Cuprolite X96DP	15 min	60°C
Uniphase PHP Pre-catalyst	15 min	20°C
Uniphase PHP Catalyst	15 min	40°C
Niplast AT78	15 min	40°C
Electroless Nickel SLOTONIP 1850	60 min	89°C

Table 2. EN co-deposition materials and procedure

A ceramic powder of 50g/l was added into the bath along with the substrate. With agitation, suspended particles near the surface are co-deposited onto the substrate surface. The pH of the EN solution was varied between the manufacturer standard pH 4.9 and pH 5.4. The pH was altered to pH 5.4 by adding 10% ammonium hydroxide. The coating time was kept constant at 60 minutes. The bath temperature was kept constant at 89 ± 2°C. The particles were kept in suspension in the EN bath by either mechanical stirring or air bubbling agitation methods. Mechanical stirring was done by Jenway hotplate with magnetic stirrer, and air bubbling was performed at1.2W air pressure.

2.3. SEM-EDX

Composition of Ni-P and YSZ composite in the deposition is controlled to give desired properties. It is desirable to get high ceramic to metal ratio for corrosion, thermal and wear resistance and even in the application of anode of fuel cell. Varying ceramic YSZ deposition in the composite has an advantage in which a gradient of coating layers with increasing ceramic content inside to outside for heat and corrosion resistance respectively. The influence of the process parameters in gaining high particle incorporation is analysed. The characterisation of the composite deposition was done by Hitachi field emission gun scanning electron microscope (SEM) coupled with energy dispersive x-ray (EDX).

The common setting of SEM is 24 mm working distance and 25 kV acceleration voltage unless stated otherwise to ensure optimum condition for EDXA. Magnification at 1000 or 2000 and resolution are varied according to the requirement (3.9-6.0). The EDX expose time was kept constant at 300s and expose area was kept constant throughout the whole specimens.

SEM enables surface morphology and chemical microanalysis in conjunction with EDX. EDX stands for Energy dispersive X-ray where an x-ray emitted during electron beam targeted to the sample surfaces is detected and collected for elemental composition characterisation. The electron beam bombardment knock-out the electron near the surface and resulting electron vacancy is filled by higher energy electron level. This energy is between 10-20 eV, depending on the materials and emits x-rays to balance the energy difference between the two electron states. EDX collaborated with INCA software extends its ability for quantitative analysis, qualitative analysis, elemental mapping, and line profile analysis.

2.4. Archimedes buoyancy

Archimedes specific density can be used to measure the porosity fraction in a material. The basic Archimedes principle states that the amount of displaced water volume is equal to the immersed object volume. The determination of the solid substance density can be done by buoyancy or displacement methods. The true density on the other hand is the total solid density. Figure 2 (Sartorius 1991) illustrates the Archimedes density buoyancy methods. The density of a solid body is defined as a product of fluid density and fraction of solid mass over fluid mass. The apparent weight of a body in a liquid – weight reduced by buoyancy force is measured. The density of water at room temperature is assumed to be unity. The substrate alumina tile is assumed to be fully dense thus the calculation of pores of deposition will not be affected by the substrate.

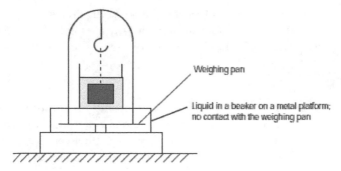

Figure 2. Schematic diagram of Archimedes density buoyancy measurement (Sartorius 1991)

2.5. Four-point probe

The composite fabricated via EN co-deposition was tested in term of its electrical resistivity and conductivity performance measured using four-point probe. The conductivity performance of the Ni-YSZ composite was tested in air environment up to 800°C. Then another set of test where the nitrogen gas was purged into the furnace up to 600°C at 20°C/min. A four-point probe was used to measure the sheet resistance and thus the resistivity and conductivity of the anode at every 5-15°C increment under 50 mA current.

The resistance of Ni-YSZ anode was measured using four-point electrical probe as illustrate in Figure 3. A power supply provides a constant current flow between probe 1 and 4. The current output can be obtained by an ammeter. The second set of probe (probe 2 and 3) is used for sensing and since negligible current flows in these probes – only voltage drop – thus accurate resistance is measured. A resistance of the sample between probes 2 and 3 is the ratio of the voltage registering on the digital voltmeter to the value of the output current of the power supply. A simple 4-point measurement at room temperature was done at 1 mA, 50 mA and 90 mA at two different points as a trial test.

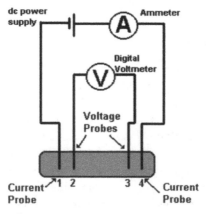

Figure 3. Schematic four-point probe set up

2.5. Full factorials

As for the EN composites process, the conventional process is disturbed by the addition of 8 mol% YSZ (8YSZ; United Ceramics, England) particles in the bath. Therefore the co-deposition process is not only affected by the basic process parameters but also other factors including the particle stability, particle size and shape, particle loading, and bath agitation (Feldstein 1990). Few process parameters were identified to be most effective in gaining the yields namely bath agitation, bath pH, particle size and surface treatment. These parameters were then design for experiment using 16 run full factorials with five replications.

The experimental design involved four EN co-deposition process parameters (Table 3). Both EN conventional and composite coating have been shown to be very much affected by bath pH, bath agitation, substrate surface condition, and by particle shape and size in the composite EN coating. A full factorials DoE approach with four parameters at two levels gave 2^4 full factorials of 16 runs. The DoE was repeated five times independently yielding 80 sets of experimental data.

Parameters	Symbols	Level	
		Low (-1)	High (+1)
Particle size	A	2	10
Bath agitation	B	Air Bubbling	Mechanical Stirring
Bath pH	C	4.9	5.4
Surface treatment	D	HF Etching	Mechanical Blasting

Table 3. EN co-deposition process parameters and their levels

3. Results and discussions

3.1. Composite characterisation

Ni-YSZ was successfully deposited on alumina substrate by EN co-deposition. The typical microstructure of EN Ni-YSZ composite deposition is shown in Figure 4. The ceramic co-deposition of 10 microns (Figure 4b) ceramic powders in EN Ni-YSZ composite is higher compared to the 2 microns (Figure 4a). The coating consists primarily of ceramic YSZ powders (white areas), metallic Ni matrix (grey areas) and pores (dark areas). In general, both coating surfaces showed uniform distribution of ceramic particles. This indicates no agglomeration of YSZ particle in the coating. Uniformly distributed ceramic particles ensure constant coefficient of thermal expansion within the coating to avoid cracking due to thermal gradient.

The corresponding EDX spectrum is given in Figure 5 shows the presence of major peak of nickel (Ni), yttria (Y), zirconium (Zr), oxygen (O) and phosphorus (P) elements with primary Kα and Lα peaks were present. These confirmed that the composites are composed of the combination of metallic nickel and ceramic YSZ. The existence of element P in the composite is due to the fact that P is one of the major elements in the EN hypophosphite-base bath solution. The Ni-P deposition reaction in hypophosphite-based bath involved the chemical reaction as in equation (1.3) described in section 1.0.

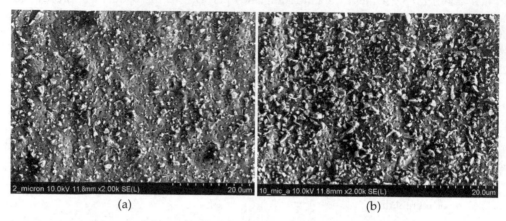

Figure 4. Comparison of surface microstructure of Ni-YSZ for ceramic particle size (a) 2 μm and (b) 10 μm

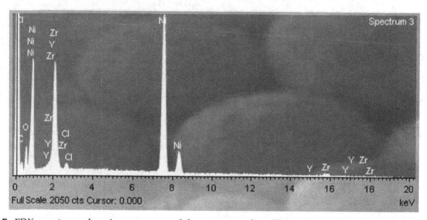

Figure 5. EDX spectrum showing presence of the corresponding EN co-deposition

In order to investigate the existence of pores at the dark areas, a higher magnification images were taken. The images show the surface morphology at various magnifications at (a) 6k, (b) 10k and (c) 20k in Figure 6. At 6k magnification image (Figure 6a) shows few dark areas which are the expected pore spots and the centre spot was focused and magnified to 10 k magnification (Figure 6b). Clearly the images confirmed the existence of pores at higher magnification of 20k (Figure 6c) the pores seemed to be open or connected.

It is expected that varying surface morphology of the substrate will increase the porosity formation. The images of field emission gun SEM at various magnifications from 1-20k are shown in Figure 7 compares the mechanical blasting treatment substrate surfaces (all images on the left represented by a, c and e) and chemical etching treatment (all images vertically on the right represented by b, d and f). At the 1k magnification, Figure 7a (blasted) indicated more highly populated black areas compared to Figure 7b (etched). The chemically etched surface morphology is much flatter.

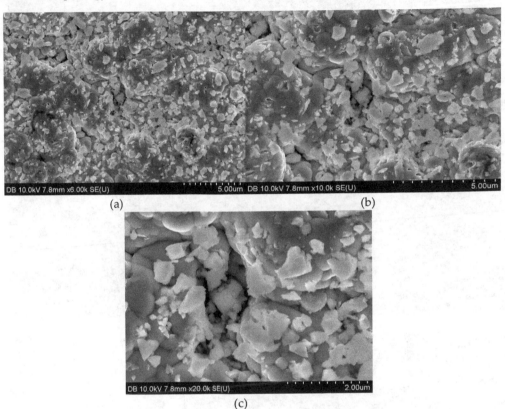

Figure 6. EN Ni- 8YSZ (2 ◉m) and bath pH5.4 at varying magnification (a) 6k (b) 10k and (c) 20k

At 5k magnification, the difference is more obvious and it is confirmed that the black areas or spots were the pores. At even higher magnification (20k), the mechanically blasted deposition (Figure 7e) exhibits many open or connected pores which are highlighted with white circles compared to the chemically etched deposition (Figure 7f).

3.2. Composite composition

Composite composition is basically an investigation of the ratio between the metallic nickel and ceramic YSZ. The composition was investigated by an analysis of the 2^k factorials design of experiment. The experiment was designed for lower nickel composition and higher ceramic incorporation in the Ni-P-YSZ composite. The main effect plot of all the four main effects namely particle size (A), agitation (B), bath pH (C) and substrate surface treatment

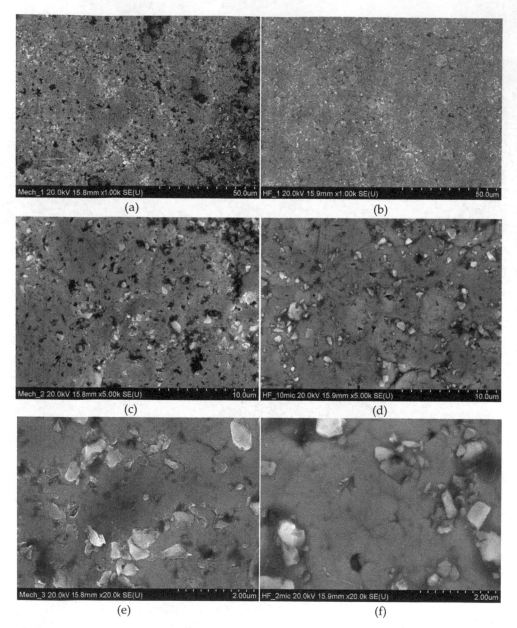

Figure 7. SEM micrographs at 1k, 5k and 20k magnification. Mechanical treated surface deposition (left: a, c, e) and chemical treated surface deposition (right: b, d, f).

(D) is illustrated in Figure 8. The main effect plot is a plot of the mean Ni content in vol.% at each level of a design parameter. The bigger the difference between the high and low levels, the higher is the effect. Referring to Figure 8, it is clearly indicated that the most significant factor is A with 3.483 strength effects followed by C (1.341), then B (0.734) and lastly D (0.036). The effect of parameter D is almost zero, since there is almost no difference between the high and low level.

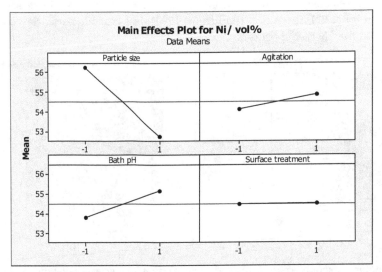

Figure 8. The main effect plots for Ni content response

Generally, by referring to the main effect plot above, particle size at high level, bath agitation at low level and bath pH at low level give the lower nickel content. This concluded that the main effects that influence the EN co-deposition process for lower Ni to YSZ deposition are particle size and bath pH. Generally, particle size (A) effect at high level (10 μm) and bath pH (C) at low level (pH 4.9) gives lower Ni to YSZ ratio.

Based on ANOVA analysis, the three-way interaction of factor A (particle size), B (bath agitation) and C (bath pH) was found to be significant. The investigation of the three-way interactions between factors A, B and C is referred to the contour plot given in Figure 9. The contour plot shows variation of the main effect factor A (particle size) and C (bath pH) where the interaction effect B (bath agitation) was kept constant at either low level of air bubbling (Figure 9a) and high level of stirring (Figure 9b). Under low level bath agitation, low Ni content was obtained at high level factor A (particle size) of 10 μm. It is also shown that factor C (bath pH) gives no effect on Ni composition at high (Figure 9a) or low level of A (Figure 9b). The low Ni content was obtained for high level bath agitation (Figure 9b) when factor A (particle size) at high level and factor C (bath pH) at low level.

The influence of factor B on both main factors A and C in determining low Ni to YSZ ratio is greatly affected by factor A (particle size). Larger particle size was found easier to be co-deposited in EN deposition rather than smaller particle size. Balaraju et. al. (2006) showed highest particle incorporation of largest alumina powder for a range of particle size between 50 nm, 0.3 μm and 1.0 μm. The lowest Ni to YSZ ratio was obtained under high level bath agitation (mechanical stirring) with Ni content less than 52 vol.%. The effect of factor C (bath pH) on Ni content showed low level gives the lowest Ni composition in both condition of factor B. The incorporation of ceramic powders is higher under the standard EN solution pH 4.9 (Schloetter 2006) ensures optimum condition. Although higher pH increased deposition rate in conventional EN deposition (Baudrand 1994), less ceramic particles were able to be incorporated in the EN co-deposition.

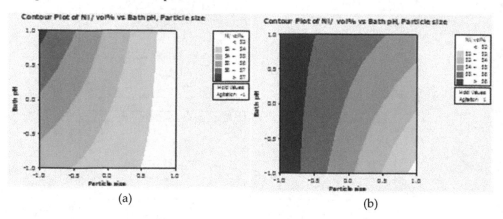

(a)   (b)

Figure 9. Contour plots for Ni content, pH and particle size at (a) low level agitation – air bubbling and (b) high level agitation – mechanical stirring

Thus to achieve low nickel to ceramic ratio, the conditions are (1) large particle size of 10 μm, (2) mechanical stirring agitation and (3) bath pH of 4.9. Larger particle size was found easier to be co-deposited in EN deposition rather than smaller particle size. This was supported by Balaraju et al (2006) on alumina powder sizes of 50 nm, 0.3 μm and 1.0 μm that resulted in the highest particle incorporation at 1.0 μm particle size. A study done by Vaghefi et al. (2003) showed 33 vol.% of B4C particle with particle size ranges 5-11 μm which indicates that larger particle sizes give higher particle incorporation in the EN composite.

In terms of bath agitation, it is crucial to keep the particles in suspension throughout the deposition (Sevugan, Selvam et al. 1993). Mechanical stirring showed higher incorporation of particles in this study compared to air bubbling. It should be noted that this is in contradiction to the finding by Vaghefi (1997) in electroless nickel-phosphorus-molybdenum disulfide which showed that air purging was better than magnetic stirring. However, the particles used in this research were ceramic YSZ which, in terms of inertness, wettability and particle stability to the substrate (Apachitei, Duszczyk et al. 1998; Necula, Apachitei et al. 2007) are different to the study by Vaghefi. As for the bath pH, it is shown that a higher bath pH caused a higher deposition rate in conventional EN deposition (Baudrand 1994). Thus at a higher deposition rate, it is possible that less ceramic particles were able to be dragged along in the EN co-deposition. This resulted in the observation that the incorporation of ceramic particle is higher at lower pH of 4.9 than at 5.4.

Therefore, the best deposition parameter combination for the three-way interaction based on this observation are high level particle size of 10 um, low level bath pH of 4.9 and high level agitation of mechanical stirring (A+1B+1C-1).

3.3. Porosity content

Investigation of porosity level in the deposition is critical as the amount of porosity enhances thermal insulation for thermal barrier coatings (Wang, Wang et al. 2011) and gas circulation in fuel cell anode (Simwonis, Thulen et al. 1999) application. The amount of porosity should not be more than 40 vol.% as greater amount of porosity will reduce the mechanical properties of the deposit. Thus, an adequate amount of porosity and reasonable mechanical properties should be balance. It is known that EN deposition has excellent uniformity and dense deposition with thickness less than 10 μm (Das and Chin 1959). The amount of porosity in EN deposition could be induced by varying the agitation methods, deposition rate, bath pH and also substrate surface condition. The porosity of deposition was measured by Archimedes density measurement.

Figure 10 shows the main effect plot indicates the variation of the data mean between low and high levels for each main parameter. The most dominant parameter as already verified by the ANOVA is the substrate surface treatment (D) with bath agitation (B) closely behind. The difference between low and high levels for both parameters D and B are the largest and difficult to differentiate. Descending rank order for the strength effect is surface treatment, bath agitation, particle size and then bath pH. The 'higher-the-better' characteristic for porosity response described that high porosity content can be achieved at low level particle size (2 μm), high level bath agitation (mechanical stirring) and high level substrate surface treatment (mechanical blasting).

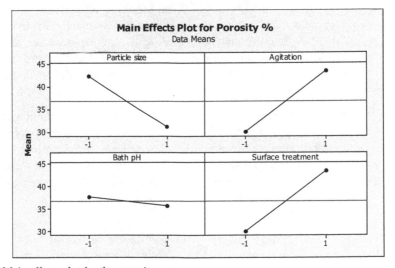

Figure 10. Main effects plot for the porosity response

The interaction plot is a powerful graphic tool which plots the mean of response of two factors. The AD interaction plot in Figure 11 shows unparallel lines of high and low level particle size under surface treatment variation. This indicates the present of interaction between the particle size and the substrate surface treatment. The porosity content was very much affected by the substrate surface treatment. Varying substrate surface treatment from low (-1) to high (+1) level increases porosity % for both particle size at low (-1) and high (+1) levels. The effect of particle size on porosity % at low level (-1) is more pronounce than at high level (+1) as the substrate surface treatment changes from low (-1) to high (+1) level. At both particle size levels, mechanical blasting gave greater effect on porosity.

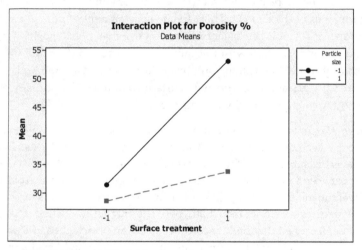

Figure 11. Figure 11: Interaction (A-D) plot for porosity

Thus to achieve high porosity content, the condition are (1) small particle size of 2 μm, (2) mechanical stirring agitation and (3) substrate surface treatment of mechanical blasting. Study by Wang et. al. (2006) has shown that fine particles introduced smaller size porosity. This indirectly indicates high porosity volume, i.e. like smaller pebbles in a jar has more quantity than the larger one. The mechanical stirring bath agitation gives higher porosity due to low removal of absorbed hydrogen or oxygen (Sevugan, Selvam et al. 1993).

The mechanical stirring bath agitation gives higher porosity as the agitation is not very aggressive compared to air bubbling thus most of the absorbed hydrogen or oxygen was not removed (Sevugan, Selvam et al. 1993) and trapped inside the EN deposition introducing more porosity. Mechanical blasting resulted in a rougher substrate surface. EN deposition is very well-known to follow the substrate profile rather than filling the spaces (Taheri, Oguocha et al. 2001). Therefore, rougher surface caused a rougher deposition surface and thus introduced more porosity.

Porosity measurement using Image Pro-Plus software was conducted. An SEM image of an EN co-deposition (8YSZ, 2 μm in bath pH 5.4) is shown in Figure 12. This image was then analysed using colour contrast for porosity measurement. The red coloured area was the amount of porosity in the deposition (Figure 12b). It can be estimated that the coloured porosity area is approximately 20%.

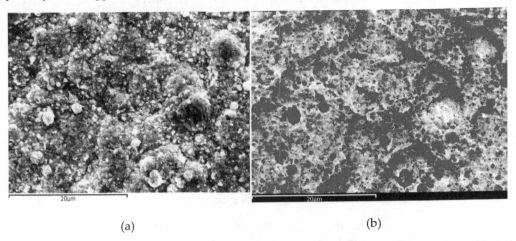

(a)   (b)

Figure 12. Image Pro-Plus porosity measurement; (a) SEM image (b) porosity area mapping

3.4. Regression modelling

A model to predict response function can be built by a regression model. The build model can be used to illustrate the relationship between the experimental data and the predicted data, describe the relationship between a response and a set of process parameters that affect the response. Other than that, the model can also be used to predict a response for different combinations of process parameters at their best levels. Based on the 80 sets of experimental data, the linear regression models were successfully developed for nickel to YSZ ratio response as shown in equation (5).

$$\hat{y} = 54.491 - 1.742A + 0.367B + 0.671C + 0.018D + 0.785ABC \cdots \qquad (5)$$

where $\hat{y}$ is the nickel content, A is particle size (μm), B is bath agitation, C is bath pH and D is surface treatment. The coefficient of determinations (R^2) was 0.72, indicating a reasonable correlation between the measured and predicted values of nickel content as shown in Figure 13. This means the model is reliable in predicting the response with 28% variation. Referring to the coefficient of the developed models, it was confirmed that particle size was the most prominent parameter in minimising the nickel to YSZ ratio.

Minitab analysis showed the optimum condition for achieving a low Ni to YSZ ratio in the coating is when factors A and B are at high level and factor C is at low level. The minimum nickel content obtained experimentally was 51.827 vol.% whereas the predicted content was 51.293 vol.%. The difference between the measured value and predicted value (0.534) is minimal, indicating that the model is reliable in predicting the response value at the desired parameter levels.

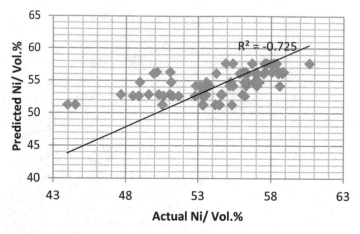

Figure 13. Comparison of experimental and predicted values of Ni content

Similarly for the Porosity %, the analysis done by Minitab in the previous section led to a reduced model that represents the optimum condition of the process in obtaining high porosity %. The reduced model equation contains main parameters A, B and D and two-way interactions of AD and is shown in equation (6).

$$\hat{y} = 36.734 - 5.614A + 6.692B + 6.776D - 4.167AD \cdots \tag{6}$$

where $\hat{y}$ is the response Porosity %, A is particle size (μm), B is bath agitation and D is substrate surface treatment. The coefficient of determinations (R^2) as shown in Figure 14 is 0.47 indicating that the correlation between the actual and predicted values of Porosity % was not very good. Referring to the coefficient of the developed model, it is confirmed that the substrate surface treatment is the most prominent parameter followed by the bath pH in increasing the Porosity %.

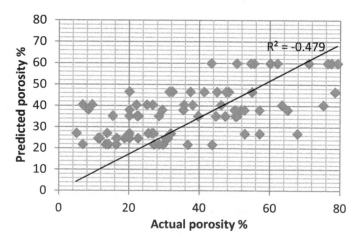

Figure 14. Comparison of experimental and predicted values for porosity content

A study by Azmir and co-authors (2009) developed the first and second order polynomial models of their four factors Taguchi giving to a reasonably high correlation between the actual and predicted values. The coefficients of regression were determined using the same software, Minitab 15. This means this approach could be applicable in this study in order to improve the correlation coefficient of these two models for future work.

3.5. Electrical conductivity

Nickel-YSZ composite is the common material for fuel cell anode and thus the electrical conductivity of this composite fabricated via EN co-deposition process is also investigated. The issue of phosphorus impurity in the EN co-deposit could impede the electronic performance of the composite. The conductivity using four-point probe both in air and nitrogen environment was measured. The air environment is to simulate the oxidation environment and nitrogen for an inert environment. The measured conductivities in different environment were found to be comparable to the published data.

The Ni-YSZ co-deposition was deposited onto a ceramic substrate representing the anode. The initial sample with the thickness of 13 μm Ni-YSZ co-deposition contains 48.32 vol.% Ni. This was co-deposited with 2 μm YSZ particle size. The initial electrical conductivity tests were carried out at room temperature (25°C) and involved measurements at two different points on the surface of the composite sample. The tests were carried out at three different currents- 1mA, 50mA and 100mA. The resistance, resistivity and conductivity of the sample at the three different currents are given in Table 4.

Current	1 mA		50 mA		100 mA	
	1st Point	2nd Point	1st Point	2nd Point	1st Point	2nd Point
Resistance/ Ω	0.117	0.210	0.353	0.425	0.317	0.261
Resistivity/ $10^{-4}\Omega cm$	1.52	2.73	4.59	5.52	4.12	3.4
Conductivity/$10^4 Scm^{-1}$	0.66	0.36	0.22	0.18	0.24	0.29

Table 4. Initial four-point electrical test at various current

The observed values were very encouraging. The initial test was carried out at room temperature using four-point probe measurement. Since YSZ is non-metallic and nickel is metallic, the Ni-YSZ composite behaves as biphasic composite system – having a conductivity percolation threshold at an adequate amount of nickel. The conductivity values of 50 vol.% Ni-YSZ is a factor of ten less than half the value of pure metallic nickel at room temperature (11.8 $x10^4$ $(\Omega cm)^{-1}$). Thus the obtained values for this initial test are still comparable.

A test on the Ni-YSZ fabricated via EN co-deposition was carried out at temperatures increasing from 25°C to 420°C in air. The scatter plot of the resistivity and conductivity are given in Figure 15. The trend of resistivity showed a linear relationship with the temperature. Resistivity increases as temperature increases. The conductivity is inversely proportional to resistivity thus it is expected to show opposite linear relationship to the resistivity. The linear decrease in conductivity with temperature is indicative of metallic conduction.

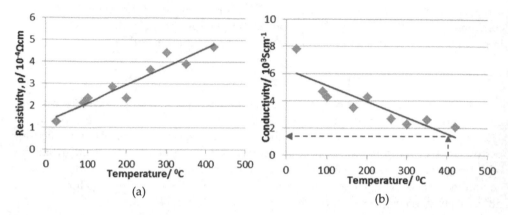

(a)   (b)

Figure 15. Scatter plot of Ni-YSZ composite (a) resistivity (b) conductivity against temperature in air

This conductivity trend was similar to the study done by Pratihar and co-authors (2005) as shown in Figure 15. The conductivity values obtained from the study at variable fabrication techniques for 40 vol.% Ni at 400°C are tabulated in Table 5. Based on the best fitted linear line, at 400°C, the conductivity value of 50 vol.% Ni is approximately 1500 Scm^{-1} (referring to the red dotted line in Figure 15b). This is comparable to the conductivity obtained by the composite fabricated via EN powder coating at 40 vol.% Ni.

Fabrication Technique	σ/ Scm^{-1}
Solid state	450
Liquid dispersion	250
EN powder coating	1100

Table 5. Conductivity values of Ni-YSZ composite fabricated by various techniques at 400°C

A study on a 50 vol.% Ni gave a conductivity of 10 Scm^{-1} (Koide, Someya et al. 2000). Comparing these values with the one obtained for Ni-YSZ composite fabricated via EN coating, the conductivity values for EN co-deposition higher by a factor of hundred. Another study for 50 vol.% Ni-YSZ by Aruna and co-authors (Aruna, Muthuraman et al. 1998) stated a value of 2.5×10^3 Scm^{-1} at 400°C which is comparable to the value obtained in this work.

Two series of electrical performance tests were conducted on another Ni-YSZ composite fabricated via EN co-deposition. These were carried out in two different environments – in air varying temperatures from 25°C to 800°C and in nitrogen varying temperatures from 25°C to 600°C. Both composite samples in air and nitrogen had coating deposition thickness of 10 microns. The conductivity plots for both series are given in Figure 16. Again, the conductivity trend decreases with temperature, an indication that it has a metallic conductivity.

The conductivity values are similar in air and nitrogen environment although the former is slightly higher. This observation might be due to the high moisture content in air compared to nitrogen. A review by Zhu and Deevi (2003) found that the Ni-YSZ composite overpotential is significantly reduced in the presence of moisture or steam. Lowering anodic overpotential enhanced the electronic conductivity.

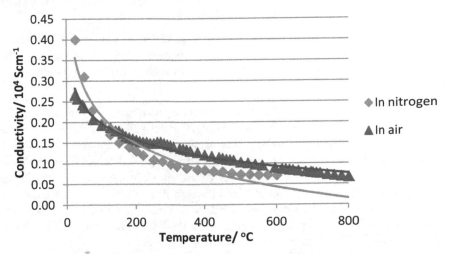

Figure 16. Scatter plot of conductivity against temperature in both nitrogen and air

In general, the conductivity at 600-800°C of Ni-YSZ fabricated via EN deposition ranged between 700-1000 Scm^{-1} in both environments. These values are comparable with the published data from several studies as tabulated in Table 6. The role of phosphorus may be important. Parkinson stated that the electrical resistivity of EN deposited nickel increases with phosphorus content (Parkinson 1997). Referring to the nickel-phosphorus phase diagram, Ni$_3$P are exists at temperatures greater than 400°C. The effect of these Ni$_3$P crystals for porosity as it could evaporate at higher temperature in fuel cell anode application should be the subject of future work. For example, this composition might be evaporated at temperature between 1107-1517°C (Viksman and Gordienko 1992).

Composite	T/°C	Fabrication	Environment	σ/ Scm^{-1}	Ref.
40vol.% Ni-YSZ	600	Solid state coating	H$_2$	1500	Kim
		Solid state mixing		900	et al. (2006)
	800	Solid state coating	H$_2$	1400	
		Solid state mixing		800	
45wt.% Ni-YSZ	600-800	Solid state with 2-step calcinations	H$_2$/Ar	500	Han et al. (2006)
		Conventional solid state mixing		430	

Table 6. Published electrical conductivity of Ni-YSZ composite

The composite fabricated via EN co-deposition is a possibility for in-situ fabrication onto ceramic substrate. It is proven the composite of 50 vol.% Ni has a metallic conductivity with highest conductivity of 1500 Scm^{-1} at 400°C in air and 700 Scm^{-1} in N$_2$; 1000 Scm^{-1} in air at 600°C and 750 Scm^{-1} at 800°C in air.

4. Conclusions

This study investigates a method of fabricating Ni-YSZ composite via EN co-deposition approach. The work involved (i) showing that successful EN co-deposition of Ni–YSZ composite by selective combinations of process parameters (ii) investigation of physical properties of the composites.

The 2^4 full factorials of 16 runs used particle sizes of 2 and 10 μm, bath agitation methods of air bubbling and mechanical stirring, bath pH of 4.9 and 5.4 and substrate surface treatment of HF etching and mechanical blasting. The design of experiment responses were nickel content and porosity content. The design of experiment were then analysed by Minitab 15 software and it was found that the optimum condition for low Ni to YSZ ratio involved a particle size of 10 μm, bath agitation of mechanical stirring, a bath pH of 4.9 and a substrate condition of HF etching. On the other hand, the porosity response optimum condition involved a particle size of 2 μm, a bath agitation method of mechanical stirring, a bath pH of 4.9 and a substrate surface treatment of mechanical blasting.

Based on the 80 sets of experimental data, the linear regression models were successfully developed for both responses. The coefficients of determinations (R^2) of nickel composition and porosity content were found to be 0.72 and 0.47 respectively. There is a reasonable correlation between the measured values and predicted value for nickel content. These models can be used in determining EN co-deposition parameters for tailored amount of nickel content.

In terms of the electrical conductivity performance, the initial electrical conductivity test carried out at room temperature showed an encouraging outcome in that the value for a 50 vol.% Ni-YSZ composite was only a factor of ten less than the equivalent loading of pure nickel. The electrical conductivity of this composite at 400°C in air was comparable to published data in other studies and was superior to those recorded for composites manufactured by traditional techniques. At temperatures up to 800°C, the electrical conductivity tests were carried out in two different environments - air and nitrogen – and results were comparable to those in the public domain.

Author details

Nor Bahiyah Baba
TATI University College (TATIUC), Terengganu, Malaysia

Acknowledgement

The author would like to express her gratitude to Mr. Alan Davidson and Prof. Tariq Muneer from *School of Engineering and Build Environment, Edinburgh Napier University, United Kingdom* for their technical support and great contribution of knowledge.

5. References

Anthony, J. (2003). *Design of Experiments for Engineers and Scientist*, Butterworth-Henimen.

Apachitei, I., J. Duszczyk, et al. (1998). "Particles Co-Deposition by Electroless Nickel." *Scripta Materialia* 38(9): 1383–1389.

Aruna, S. T., M. Muthuraman, et al. (1998). "Synthesis and properties of Ni-YSZ cermet: anode materials for solid oxide fuel cells." *Solid State Ionics* 111: 45-51.

Azmir, M. A., A. K. Ahsan, et al. (2009). "Effect of abrasive water jet machining parameters on aramid fibre reinforced plastics composite." *International Journal Materials Form* 2: 37-44.

Baba, N. B., W. Waugh, et al. (2009). Manufacture of Electroless Nickel/YSZ Composite Coatings. *World Congress of Science, Engineering and Technology (WCSET)* Dubai, UAE, WASET 2009 ISSN 2070-3740, Vol.37. ISSN 2070-3740, Vol.37: 715-720.

Balaraju, J. N., Kalavati, et al. (2006). "Influence of particle size on the microstructure, hardness and corrosion resistance of electroless Ni–P–Al2O3 composite coatings." *Surface & Coatings Technology* 200: 3933 – 3941.

Balaraju, J. N., T. S. N. S. Narayanan, et al. (2006). "Structure and phase transformation behaviour of electroless Ni–P composite coatings." *Materials Research Bulletin* 41: 847–860.

Balaraju, J. N. and K. S. Rajam (2008). "Preparation and characterisation of sutocatalytic low phosphorus nickel coatings containing submicron nitride particles." *Journal of Alloys and Compounds* 459(1-2): 311–319.

Baudrand, D. W. (1994). "Electroless Nickel Plating " *ASM Handbook Volume 5* Surface Engineering: 290-310.

Berkh, S. Eskin, et al. (March 1996). Properties of Electrodeposited NiP-SiC Composite Coatings. *Metal Finishing*: 35-40.

Bisgaard, S. (1998). A Practical Aid for Experimenters. *Preliminary Edition*. Madison, Starlight Press: 57.

Dai, J., X. Liu, et al. (2009). "Preparation of Ni-coated Si3N4 powders via electroless plating method." *Ceramics International* 35(8): 3407–3410.

Das, C. M., P. K. Limaye, et al. (2007). "Preparation and characterization of silicon nitride codeposited electroless nickel composite coatings." *Journal of Alloys and Compounds* 436: 328-334.

Das, L. and D. T. Chin (1959). "Electrochemical Porosity Measurement of EN Coating." *Plating and Surface Finishing* 84: 66-68.

Dong, D., X. H. Chen, et al. (2009). "Preparation and properties of electroless Ni-P-SiO2 composite coatings." *Applied Surface Science* 255: 7051-7055.

Feldstein, N. (1990). Composites Electroless Plating. *Electroless Plating: Fundamentals and Applications*. G. O. Mallory and J. B. Hajdy. Orlondo, FI, AESF Publication. Chapter 11: 269-287.

Gengler, J. J., C. Muratore, et al. (2010). "Yttria-stabilized zirconia-based composites with adaptive thermal conductivity " *Composites Science and Technology* 70(14): 2117-2122.

Ger, M.-D. and B. J. Hwang (2002). "Effect of surfactants on codeposition of PTFE particles with electroless Ni-P coating." *Materials Chemistry and Physics* 76(1): 38-45.

Han, K. R., Y. Jeong, et al. (2006). "Fabrication of NiO/YSZ anode material for SOFC via mixed NiO precursors." *Materials Letters* 61: 1242–1245.

Hazan, Y. d., T. Reutera, et al. (2008). "Interactions and dispersion stability of aluminum oxide colloidal particles in electroless nickel solutions in the presence of comb polyelectrolytes." *Journal of Colloid and Interface Science* 323: 293–300.

Hazan, Y. d., D. Werner, et al. (2008). "Homogeneous Ni-P/Al2O3 nanocomposite coatings from stable dispersions in electroless nickel baths." *Journal of Colloid and Interface Science* 328: 103–109.

Huang, Y. S., X. T. Zeng, et al. (2003). "Development of electroless Ni-P-PTFE-SiC composite coating." *Surface & Coating Technology* 167: 207-211.

Hung, C. C., C. C. Lin, et al. (2008). "Tribological studies of electroless nickel/ diamond composite coatings on steels." *Diamond & Related Materials* 17: 853–859.

Kalantary, M. R., K. A. Holbrook, et al. (1993). "Optimisation of a Bath for Electroless Plating and its use for the Production of Ni-P-SiC Coatings." *Trans. Inst. Metal Finish* 71(2): 55-61.

Kim, S.-D., H. Moon, et al. (2006). "Performance and durability of Ni-coated YSZ anodes for intermediate temperature solid oxide fuel cells." *Solid State Ionics* 177: 931-938.

Koide, H., Y. Someya, et al. (2000). "Properties of Ni/YSZ cermet as anode for SOFC." *Solid State Ionics* 132: 253-260.

Lekka, M., C. Zanella, et al. (2010). "Scaling-up of the electrodeposition process of nano-composite coating for corrosion and wear protection " *Electrochimica Acta* 55(27): 7876-7883.

Li, L., M. An, et al. (2005). "Model of electroless Ni deposition on SiC$_P$/Al composites and study of the interfacial interaction of coatings with substrate surface." *Applied Surface Science* 252: 959-965.

Li, L., M. An, et al. (2006). "A new electroless nickel deposition technique to metallise SiC$_P$/Al composites." *Surface & Coatings Technology* 200: 5102 – 5112.

Li, L. B., M. Z. An, et al. (2005). "Electroless deposition of nickel on the surface of silicon carbide/aluminum composites in alkaline bath." *Materials Chemistry and Physics* 94: 159-164.

Li, Y. (1997). Investigation of Electroless Ni-P-SiC Composite Coatings. *Plating & Surface Finishing.* November 1997: 77-81.

Lin, C. J., K. C. Chen, et al. (2006). "The cavitation erosion behavior of electroless Ni–P–SiC composite coating." *Wear* 261: 1390–1396.

Liu, W. L., S. H. Hsieh, et al. (2006). "Temperature and pH dependence of the Electroless Ni-P deposition on Silicon." *Thin Solid Films* 510: 102 – 106.

Matsubara, H., Y. Abe, et al. (2007). "Co-deposition mechanisms of nano-diamond with electrolessly plated nickel films." *Electrochimica Acta* 52: 3047-3052.

McCormack, A. G., M. J. Pomeroy, et al. (2003). *Journal of Electrochemical Society* 150 (5): C356-C361.

Necula, B. S., I. Apachitei, et al. (2007). "Stability of nano-/microsized particles in deionized water and electroless nickel solutions." *Journal of Colloid and Interface Science* 314: 514–522.

Parkinson, R. (1997). Properties and Application of Electroless Nickel, Nickel Development Institute: 33.

Periene, N., A. Cesuniene, et al. (1994). Codeposition of Mixtures of Dispersed Particles With Nickel-Phosphorus Electrodeposits. *Plating ans Surface Finishing*. October 1994: 68-71.

Pratihar, S. K., A. D. Sharma, et al. (2005). "Processing microstructure property correlation of porous Ni–YSZ cermets anode for SOFC application." *Materials Research Bulletin* 40: 1936–1944.

Pratihar, S. K., A. D. Sharma, et al. (2007). "Properties of Ni/YSZ porous cermets prepared by electroless coating technique for SOFC anode application." *Journal of Materials Science* 42: 7220-7226.

Rabizadeh, T. and S. R. Allahkaram (2011). "Corrosion resistance enhancement of Ni–P electroless coatings by incorporation of nano-SiO_2 particles." *Materials and Design* 32(1): 133-138.

Sartorius (1991). Manual of weighing applications Part 1: Density: 1-62.

Schloetter (2006). Electroless Nickel - Solotonip 1850. *Bath 18810 - PE*. Worchestershire, England: 1-11.

Sevugan, K., M. Selvam, et al. (1993). Effect of Agitation in Electroless Nickel Deposition. *Plating and Surface Finishing*: 56-58.

Sharma, S. B., R. C. Agarwala, et al. (2005). "Development of Electroless Composite Coatings by using in-situ Co-deposition followed by co-deposition process." *Sadhana* 28: 475–493.

Sheela, G. and M. Pushpavanam (2002). Diamond-dispersed electroless nickel coatings. *Metal finishing*. January 2002: 45-47.

Shibli, S. M. A., V. S. Dilimon, et al. (2006). "ZrO_2-reinforced Ni–P plate: An effective catalytic surface for hydrogen evolution." *Applied Surface Science* 253: 2189–2195.

Simwonis, D., H. Thulen, et al. (1999). "Properties of Ni/YSZ porous cermets for SOFC anode substrates prepared by tape casting and coat-mix process." *Journal of Materials Processing Technology* 92-93: 107-111.

Taheri, R., I. N. A. Oguocha, et al. (2001). "The tribological characteristics of electroless NiP coatings." *Wear* 249: 389–396.

Teixeira, L. A. C. and M. C. Santini (2005). "Surface Conditioning of ABS for Metallisation into the use of Chromium baths." *Journal of Materials Processing Technology* 170: 37–41.

Vaghefi, S. M. M., A. Saatchi, et al. (2003). "Deposition and properties of electroless Ni-P-B_4C composite coatings." *Surface and Coatings Technology* 168: 259-262.

Vaghefi, S. M. M., A. Saatchi, et al. (1997). "The effect of agitation on electroless nickel-phosphorus-molybdenum disulfide composite plating." *Metal Finishing* 95(6): 102.

Viksman, G. S. and S. P. Gordienko (1992). "Behavior in vacuum at high temperatures and thermodynamic properties of nickel phosphide Ni_3P." *Poroshkovskaya Metallurgiya* 12(360): 70-72.

Wang, L., Y. Wang, et al. (2011). "Influence of pores on the thermal insulation behavior of thermal barrier coatings prepared by atmospheric plasma spray " *Materials and Design* 32(1): 36-47.

Wang, Y., M. E. Walter, et al. (2006). "Effects of powder sizes and reduction parameters on the strength of Ni–YSZ anodes." *Solid State Ionics* 177: 1517-1527.

Wu, Y., B. Shen, et al. (2006). "The tribological behaviour of electroless Ni–P–Gr–SiC composite." *Wear* 261: 201–207.

Zhu, W. Z. and S. C. Deevi (2003). "A review on the status of anode materials for solid oxide fuel cells." *Materials Science and Engineering* A362: 228-239.

Zuleta, A. A., O. A. Galvis, et al. (2009). "Preparation and characterization of electroless Ni–P–Fe$_3$O$_4$ composite coatings and evaluation of its high temperature oxidation behaviour." *Surface & Coatings Technology* 203: 3569–3578.

Chapter 9

Carbon Nanotube Reinforced Alumina Composite Materials

Go Yamamoto and Toshiyuki Hashida

Additional information is available at the end of the chapter

1. Introduction

Novel materials and processing routes provide opportunities for the production of advanced high performance structures for different applications. Ceramic matrix composites are one of these promising materials. Engineering ceramics such as Al_2O_3, Si_3N_4, SiC and ZrO_2 produced by conventional manufacturing technology have high stiffness, excellent thermostability and relatively low density, but extreme brittle nature restricted them from many structural applications (Mukerji, 1993). Considerable attention has been adopted to improve the fracture toughness. An approach has been paid to the development of nanocrystalline ceramics with improved fracture properties. Decreasing the grain size of ceramics to the sub- and nano-meter scale leads to a marked increase in fracture strength (Miyahara et al., 1994). However, fracture toughness of nanocrystalline ceramics generally displays modest improvement or even deterioration (Miyahara et al., 1994; Rice, 1996; Yao et al., 2011). As one possible approach, incorporation of particulates, flakes and short/long fibers into ceramics matrix, as a second phase, to produce tougher ceramic materials is an eminent practice for decades (Evans, 1990). Recently, researchers have focused on the carbon nanomaterials, in particular carbon nanotubes (CNTs), which are nanometer-sized tubes of single- (SWCNTs) or multi- layer graphene (MWCNTs) with outstanding mechanical, chemical and electrical properties (Dai et al., 1996; Ebbesen et al., 1996; Treacy et al., 1996; Huang et al., 2006; Peng et al., 2008), motivating their use in ceramic composite materials as a fibrous reinforcing agent.

It is well recognized that some difficulties appear to be the major cause for the limited improvement in CNT/ceramic composites prepared to date. The first is the inhomogeneous dispersion of CNTs in the ceramic matrix. Pristine CNTs are well known for poor solubilization, which leads to phase segregation in the composite owing to the van der Waals attractive force (Chen et al., 1998). Such clustering produces a negative effect on the physical and mechanical properties of the resultant composites (Yamamoto et al., 2008). The second is the difficulty in controlling connectivity between CNTs and the ceramics matrix, which leads to a limited stress transfer capability from the matrix to the CNTs (Peigney, 2003; Sheldon & Curtin, 2004; Chen et al., 2011). The strengthening and toughening mechanisms of composites by fibers are now well established (Evans, 1990; Hull & Clyne, 1996); central to an understanding is the concept of interaction between the matrix and reinforcing phase during the fracture of the composite. The fracture properties of such composites are dominated by the fiber bridging force resulting from debonding and sliding resistance, which dictates the major contribution to the strength and toughness. Thus, the adequate connectivity with the matrix, and uniform distribution within the matrix are essential structural requirements for the stronger and tougher CNT/ceramic composites. To overcome these obstacles, various efforts, such as surface modification (De Andrade et al., 2008; Yamamoto et al., 2008; Kita et al., 2010; Gonzalez-Julian et al., 2011), heterocoagulation (Fan et al., 2006a, 2006b), extrusion (Peigney et al., 2002), and their combination, have been made to effectively achieve good dispersion of CNTs in ceramic matrix. Until now, however, most results for strengthening and toughening have been disappointing, and only little or no improvement have been reported in CNT/ceramic composite materials, presumably owing to the difficulties in homogeneous dispersion of CNTs in the matrix and in formation of adequate interfacial connectivity between two phases.

This chapter presents that novel processing approach based on the precursor method. The MWCNTs used in this study are modified with an acid treatment. Combined with a mechanical interlock induced by the chemically modified MWCNTs, this approach leads to improved mechanical properties. Mechanical measurements on the composites revealed that only 0.9 vol.% acid-treated MWCNT addition results in 37% and 36% simultaneous increases in bending strength (689.6 ± 29.1 MPa) and fracture toughness (5.90 ± 0.27 MPa·m$^{1/2}$), respectively, compared with a MWCNT-free alumina sample prepared under similar processing conditions. Structure-property relationship of present composites will be explained on the basis of the detailed nano/microstructure and fractographic analysis. We also explain why previous reports indicated only modest improvements in the fracture properties of MWCNT based ceramic composites. Here, the failure mechanism of the MWCNTs during crack opening in a MWCNT/alumina composite is investigated through transmission electron microscope (TEM) observations and single nanotube pullout tests. Achieving tougher ceramic composites with MWCNTs is discussed based on these results.

2. A novel approach for preparation of MWCNT/alumina composites

To disperse the MWCNTs homogeneously in the matrix and improve the connectivity between MWCNTs and matrix, we developed a novel approach with combination of a precursor method for synthesis of an alumina matrix, an acid treatment of MWCNTs and a spark plasma sintering method. The improvement on the bending strength and fracture toughness was confirmed by the fracture tests.

2.1. Materials and specimen preparation

2.1.1. Starting materials

The MWCNT material (Nano Carbon Technologies) used in this research was synthesized by a catalytic chemical vapor deposition method followed by high temperature annealing at 2600°C. The purity was claimed to be 99.5% by the producer. Fig. 1 shows scanning electron microscope (SEM, Hitachi S-4300) and TEM (Hitachi HF-2000) images of the pristine MWCNTs. It can be seen from Fig. 1b that the pristine MWCNTs have a highly crystalline multi-walled structure with a narrow central channel. The corresponding geometrical and mechanical properties of the pristine MWCNT are listed in Table 1. The estimated diameter and length of the pristine MWCNTs from SEM and TEM measurements ranged from 33 to 124 nm (average: 70 nm) and 1.1 to 22.5 μm (average: 8.7 μm), respectively. Tensile-loading experiments with individual MWCNTs using a nanomanipulator tool operated inside SEM revealed that the tensile strengths of 10 pristine MWCNTs ranged from ~2 to ~48 GPa (average: 20 GPa) and the Young's modulus ranged from ~50 to ~1360 GPa (average: 790 GPa) (Yamamoto et al., 2010). It seems that the average tensile strength of the pristine MWCNT used in this research were somewhat lower than that of the arc-discharge grown MWCNTs (Yu et al., 2000).

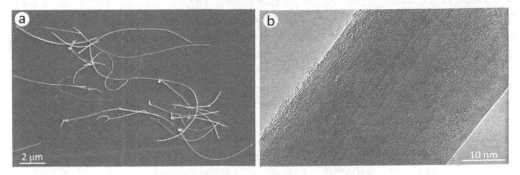

Figure 1. (a) SEM and (b) TEM image of pristine MWCNTs used in this research.

ID (nm)	OD (nm)	l (μm)	ρ (Mg/m³)	σ_f (GPa)	E (GPa)	I (nm⁴)	EI (N·nm²)
7 (3~12)	70 (33~124)	8.7 (1.1~22.5)	2.1	20 (2~48)	790 (50~1360)	1.2×10^6	9310×10^{-4}

Table 1. Measured geometrical and mechanical properties of pristine MWCNTs. Shown are the nanotube inner diameter (ID), outer diameter (OD), length (l), density (ρ), tensile strength (σ_i), Young's modulus (E), moment of inertia of cross sectional area (I) and flexural rigidity (EI), respectively.

2.1.2. Acid treatment of MWCNTs

The rationale behind the acid treatment is to introduce nanoscale defects and adsorb negatively charged functional groups at the MWCNT ends and along their lengths. The pristine MWCNTs were refluxed in 3:1 (volume ratio) concentrated H_2SO_4:HNO_3 mixture at a temperature of 70°C for 1 hour, 2 hours and 4 hours, washed thoroughly with distilled water to be acid-free, and then finally dried in an air oven at 60°C. Fig. 2 shows the typical TEM images of a series of the acid-treated MWCNTs and the corresponding distribution of nanodefect depths treated with the various conditions. It is demonstrated that with the acid treatment of the pristine MWCNTs, we have deliberately introduced nanoscale defects on the surface of the MWCNTs. The depth of the nanodefects is on the nanoscale and the average size is in the range of 4.4~7.0 nm for the acid treatment used in this study. We can see that the nanodefects density, i.e., the number of nanodefects per unit of a MWCNT surface area increases with the increasing treatment time. Hereafter, the number of the nanodefects per unit of the MWCNT surface area is referred to as the nanodefect density.

In addition to the nanodefects density, the average size of the nanodefect depths appears to vary with respect to the treatment time. When the treatment time increases from 1 hour to 2 hours, the average size of nanodefect depths increases from 4.4 nm to 6.5 nm. Furthermore, when the treatment time further increases to 4 hours, it increases to 7.0 nm. The aspect ratio (α) of the nanodefects on the MWCNT surface were estimated using the equation $\alpha = L_{width}/L_{depth}$, where L_{width} is the average size of nanodefect widths on the MWCNT surface, and L_{depth} is the average size of nanodefect depths. For the acid treated products with treatment time of 1 hour, 2 hours and 4 hours was 4.4, 4.9 and 3.9, respectively. The experimental results demonstrate that the present method, which uses the acid treatment, may provide an effective route for preparation of the nanodefects on the MWCNT surface, and it may be possible to adjust and control the average size and nanodefect density by varying the treatment time. According to the current TEM observations, peel-off of a few layers in the MWCNT structure was frequently observed for the MWCNT powders acid-treated for 4 hours. Thus, reduction of α may be due to the decrease in the MWCNT diameter by the peel-off of a few layers in the MWCNT structure and imply that the excessive acid treatment of the MWCNT resulted in degradation of the quality and mechanical properties of MWCNTs (Yamamoto et al., 2010). As previously reported (Liu et al., 1998), SWCNTs can be cut into shorter segments by acid treatment of 3:1 (volume ratio) concentrated H_2SO_4:HNO_3 mixture. In this study, however, when the acid treatment times are 1 hour and 2 hours, no such change in the length has been found in the acid-treated MWCNTs. The average lengths of the acid-treated MWCNTs were 8.7 μm and 8.3 μm, respectively. In contrast, average length was decreased slightly with a further increase in the treatment time up to 4 hours, and reached about 7.2 μm.

The zeta potential values of the pristine MWCNTs and the acid-treated MWCNTs at different pH values are shown in Fig. 3a. Here, the changes in zeta potentials were measured in 1.0 mM KCl aqueous solution of varying pH using a zeta potential analyzer (ZEECOM ZC2000, Microtec). The pH value of the aqueous solution was adjusted with HCl and NaOH. Zeta potential values were calculated using the Smoluchowski equation. The isoelectric point (pH_{iep}) for the pristine MWCNTs is located at about 3.0, whereas the acid treatment process makes the surface more negatively charged at tested pH values. The change in the zeta potential may be mainly due to the introduction of more functional groups after the acid treatment (Esumi et al., 1996; Liu et al., 1998). These functional groups make them easily dispersed in polar solvents, such as water and ethanol. Fig. 3b shows a photograph of the pristine MWCNTs and acid-treated MWCNTs suspensions at pH 6, respectively. It is clear that the pristine MWCNTs are not dispersed at pH 6. In contrast, the dispersion of the acid-treated MWCNTs is seen to improve dramatically. Furthermore, it can be expected that the larger electrical repulsive force between the acid-treated MWCNTs will facilitate their dispersion and prevent them from tangling and agglomeration. The zeta potential of the aluminum hydroxide, which is used as the starting material for synthesis of the alumina matrix, exhibited positive values over a wide pH range (pH = 3~9), while that of the acid-treated MWCNTs was negative in this pH range. On these two colloidal suspensions are mixed, particles of the aluminum hydroxide will bind onto the acid-treated MWCNTs because of the strong electrostatic attractive force between them, and this results in a homogeneous MWCNTs and aluminum hydroxide solution.

2.2. Preparation of MWCNT/alumina composites

A typical synthesis procedure for the composite preparation is as follows. The 50 mg MWCNTs acid-treated for 2 hours or pristine MWCNTs were dispersed in 400 ml ethanol with aid of ultrasonic agitation. 15.2 g aluminum hydroxide (Wako Pure Chemical Industries) was added to this solution and ultrasonically agitated. 73 mg magnesium hydroxide (Wako Pure Chemical Industries) was added to prevent excessive crystal growth. Here, the weight loss of the hydroxides caused by the dehydration process was accounted for in the calculation of the composite composition. The weight loss of the aluminum hydroxide and the magnesium hydroxide was 34.7% and 31.9%, respectively. The resultant suspension was filtered and dried in an air oven at 60°C. Finally, the product obtained in the previous step was put into a half-quartz tube and was dehydrated at 600°C for 15 min in argon atmosphere. The composites were prepared by spark plasma sintering (SPS, SPS-1050 Sumitomo Coal Mining) (Omori, 2000) in a graphite die with an inner diameter of 30 mm at a temperature of 1500°C under a pressure of 20 MPa in vacuum for 10 min. For comparison, similar preparation processes were applied while using the pristine MWCNTs as the starting material. Fig. 4 shows X-ray diffraction patterns (M21Mac Science) of the (a) aluminum hydroxide–MWCNT mixture, (b) dehydrated product and (c) sintered body, respectively. It is difficult to distinguish the MWCNT peaks from all XRD patterns, probably due to the small quantity of MWCNTs. The diffraction peaks corresponding to the aluminum hydroxide and the intermediate were observed in the aluminum hydroxide–MWCNT mixture. However, the diffraction peaks corresponding to the aluminum hydroxide and the intermediate disappeared completely in the sintered body, suggesting the phase transformation of the aluminum hydroxide to form α-alumina via an amorphous phase (b). These results clearly indicate that alumina was successfully synthesized by SPS at 1500°C under 20 MPa in vacuum.

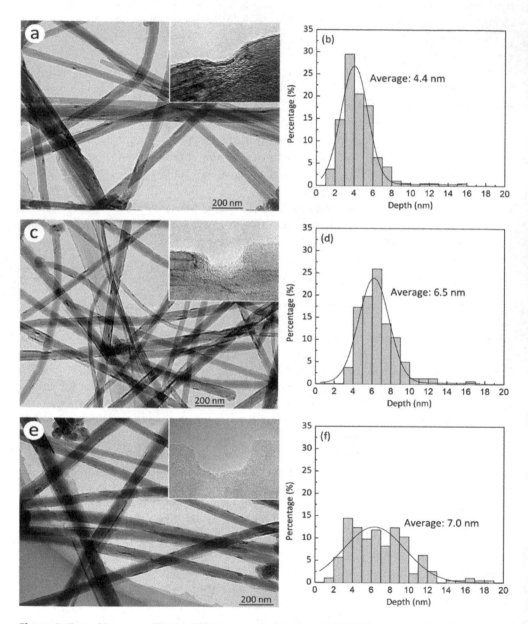

Figure 2. Typical low-magnification TEM images of acid-treated MWCNTs treated with the various conditions of (a) 1 hour, (c) 2 hours and (e) 4 hours. The insets show the high magnification images. (b), (d), (f) Corresponding depth distribution of the nanodefects in the sample (a), (c) and (e), respectively. The solid lines in (b), (d), (f) represent the Gaussian fitting curves. The observation was made for approximately 200 defects.

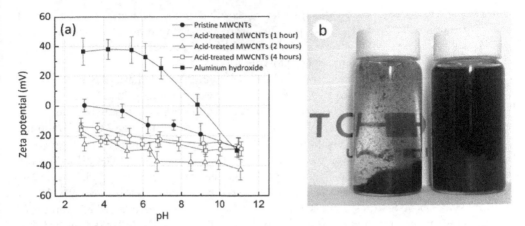

Figure 3. (a) Zeta potential values of MWCNTs and acid-treated MWCNTs at different pH. (b) Undistributed one-day old aqueous suspensions of (left) pristine MWCNTs and (right) acid-treated MWCNTs.

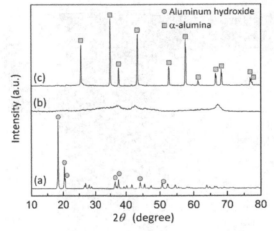

Figure 4. XRD patterns of the (a) aluminum hydroxide–MWCNT mixture, (b) dehydrated product and (c) sintered body, respectively.

2.3. Micro- and nanostructures of MWCNT/alumina composites

We now discuss the micro- and nanostructures of the acid-treated MWCNT/alumina composites using SEM and TEM analysis. An interesting geometric structure was observed between the individual MWCNT and the alumina matrix, as shown in Fig. 5. It is revealed that a nanodefect on the acid-treated MWCNT is filled up with alumina crystal, which may be intruding into the nanodefect during grain growth. This nanostructure is novel in that its structure resembles a nanoscale anchor with an alumina crystal spiking the surface of the MWCNT.

From the SEM observations on the fracture surface, the following features can be noted. First, numerous individual MWCNTs protrude from the fracture surface, and the pullout of the MWCNTs can be clearly observed (Fig. 6a), which had not been obtained until now for conventional CNT/ceramic composites. Most of MWCNTs are located in the intergranular phase and their lengths are in the range 0~10 μm. The alumina grains have sizes in the micron range, around 1.5 μm (The grain size of the composite was obtained using SEM images, and the observation was made for 224 grains.). No clear difference in the grain size is observed between the acid-treated MWCNT/alumina composites and the pristine MWCNT/alumina composites, even though the incorporation of MWCNTs seems to suppress the grain growth of the alumina. Second, in the case of the smaller amount of the acid-treated MWCNTs, no severe phase segregation was observed, whereas the composites made with the pristine MWCNTs revealed an inhomogeneous structure even for MWCNT addition as low as 0.9 vol.%. In addition to the above features, some MWCNTs on the fracture surface showed a "clean break" near the crack plane, and that the diameter of MWCNT drastically slenderized toward their tip, as illustrated in Figs. 6b and 6c, respectively. As SEM cannot clearly resolve the thickness of a single MWCNT, TEM was used to determine if the fracture phenomenon of MWCNTs was indeed occurring during crack opening.

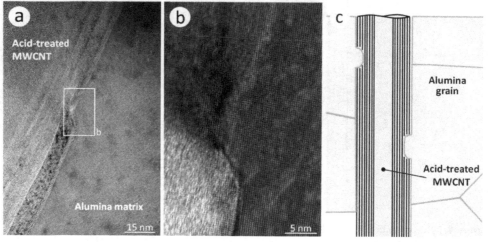

Figure 5. MWCNT morphology in the composites. (a) It is demonstrated that a nanodefect on the acid-treated MWCNT is filled up with alumina crystal. (b) Enlarged TEM image, taken from the square area. (c) Schematic description of MWCNT morphology in the composites.

TEM observations on the fracture surface demonstrated that a diameter change in the MWCNT structure was evidently observed for a certain percentage of the MWCNTs (Fig. 7a). At least, 25% MWCNT appear to have an apparent diameter change (The observation was made for 281 MWCNTs.). As shown in Fig. 7b, the high magnification TEM image clearly showed a change in diameter, and this morphology is quite similar to a "sword-in-sheath"-type failure (Yu et al., 2000; Peng et al., 2008; Yamamoto et al., 2010). Key features are illustrated in enlarged TEM image, taken from the square area in Fig. 7b. The inset showed that outer-walls having approximately 10 shells were observed to break up at the location where the MWCNT undergo failure, and that the edges of the broken outer shells were observed to be perpendicular to the cylinder axis. Since no apparent variation in the diameter of the MWCNTs has been observed along the axis in the as-received MWCNTs, these results imply that some MWCNTs underwent failure in the sword-in-sheath manner prior to pullout from the matrix. Note that MWCNT failure was also observed in fracture surfaces of alumina composites made with arc-discharge-grown and chemical vapor deposition-grown MWCNTs prepared under the same processing conditions (Yamamoto et al., 2011).

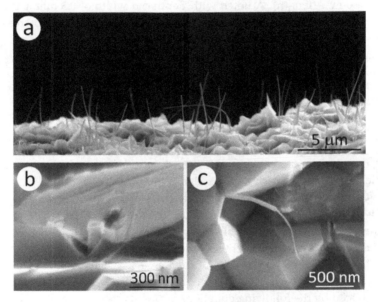

Figure 6. Fracture surface of acid-treated MWCNT/alumina composites. (a) Numerous individual MWCNTs protrude from the fracture surface. (b,c) Some MWCNTs have broken in the multi-wall failure.

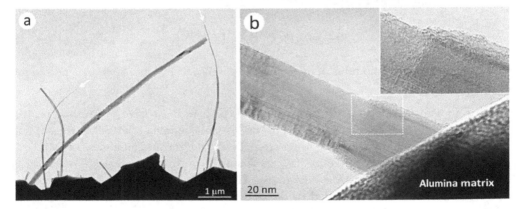

Figure 7. TEM images of the fracture surface of the composite acquired (a) low and (b) high magnification images.

2.4. Physical and mechanical properties of MWCNT/alumina composites

The bending strength of the composites was measured by the three-point bending method under ambient conditions, in which the size of the test specimens was 2.0 mm (width) × 3.0 mm (thickness) × 24.0 mm (length). The span length and crosshead speed for the strength tests were 20.0 mm and 0.83 μm/s, respectively. The fracture toughness was measured by the single-edge notched beam (SENB) method (Japanese Industrial Standards, 1995) under ambient conditions, in which the size of test specimens was 2.0 mm (width) × 3.0 mm (thickness) × 15.0 mm (length). A notch with depth and width of 0.3 mm and 0.1 mm was cut in the center part of the test specimens. A span length of 12.0 mm and crosshead speed of 0.83 μm/s were applied for the toughness test. The bending strength (σ_b) and fracture toughness (K_{Ic}) are given by the following equations:

$$\sigma_b = 3P_b L / 2bh^2 \tag{1}$$

$$K_{Ic} = \left(3P_b L / 2bh^2 \right) \cdot a^{1/2} Y \tag{2}$$

where P_b is the maximum load, L is the span length, b is the specimen width, h is the specimen thickness, a is the notch depth and Y is the dimensional factor. All surfaces of the specimens were finely ground on a diamond wheel, and the edges were chamfered. The indentation tests were done on a hardness tester (AVK-A, Akashi) with a diamond Vickers indenter under ambient conditions. The 0.9 vol.% acid-treated MWCNT/alumina composite with surface roughness of 0.1 μm (Ra) was indented using a Vickers diamond pyramid with a load of 98.1 N (P) applied on the surface for 15 s. The diagonal (d) and the radial crack length (C) were measured by the SEM. The hardness (Hv) and indentation toughness values (K_{Ic}) were calculated by the following equations:

$$Hv = 0.1891P / d^2 \tag{3}$$

$$K_{Ic} = 0.016 \left(E / H_v \right)^{1/2} \left(P / C^{3/2} \right) \tag{4}$$

where E is the Young's modulus of the composite ($E = 362.8$ GPa) measured by a pulse-echo method.

It was found that surface modification of the MWCNTs is effective in improvement of bending strength and fracture toughness of the MWCNT/alumina composites. Figs. 8a and 8b show the dependence of the bending strength and the fracture toughness on MWCNT content in the composites. There are few papers which report significant improvement in the mechanical properties such as toughness (Zhan et al., 2003), and the improvement by MWCNT addition has been limited so far in previous studies (Ma et al., 1998; Sun et al., 2002; Wang et al., 2004; Sun et al., 2005; Cho et al., 2009). In our composites, however, the bending strength and the fracture toughness simultaneously increased with the addition of a small amount of the acid-treated MWCNTs. The bending strength and the fracture toughness of the 0.9 vol.% acid-treated MWCNT/alumina composite reached 689.6 ± 29.1 MPa and 5.90 ± 0.27 MPa·m$^{1/2}$, respectively. At the same time, the bending strength and the fracture toughness of the acid-treated MWCNT/alumina composites were always higher than those of the pristine MWCNT/alumina composites with identical MWCNT content, indicating enhanced stress transfer capability from the alumina to the acid-treated MWCNTs. The Vickers indentation toughness calculated by using the Eq. (4) was 6.64 MPa· m$^{1/2}$, which is a slightly larger value than that measured by using SENB method (5.90 MPa· m$^{1/2}$). These observations revealed that the high structural homogeneity and enhanced frictional resistance of the structural components led to a simultaneous increase in the strength and the toughness of the acid-treated MWCNT/alumina composites. In contrast, for the larger amount of the MWCNTs, the degradation of mechanical properties of both the composites may be primarily attributed to the severe phase segregation. Because a bundle of segregated CNTs has poor load-carrying ability, the effect of this kind of CNT aggregate in the matrix may be similar to that of pores (Yamamoto et al., 2008a, 2008b).

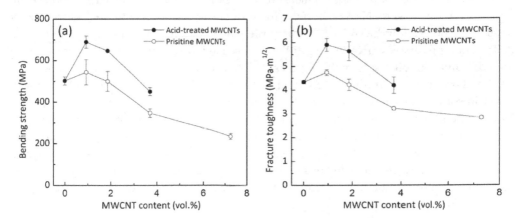

Figure 8. (a) Bending strength and (b) fracture toughness as a function of MWCNT content.

3. Evaluation of crack bridging characteristics

Ceramic-CNT interfacial behavior is another key factor in controlling the mechanical and physical properties of fiber reinforced composite materials (Evans, 1990; Hull & Clyne, 1996; Chen et al., 2011). In general, strong interfacial connectivity facilitates effective load transfer effect, but it prevents CNT pull-out toughening from occurring. Weak interfacial connectivity favors CNTs pull-out but fails to toughen the ceramic matrix. Thus, a balance must be maintained between CNT pull-out and toughening mechanics. It is well recognized that improved toughness of fiber-reinforced ceramic composites is obtained under moderate fiber-ceramic interfacial connectivity. In this regard, suitable (neither too strong nor too weak) ceramic-CNT interfacial connectivity is needed to ensure effective load transfer, and to enhance the toughness and strength of ceramic-CNT composites.

Here, the failure mechanism of the MWCNTs during crack opening in a MWCNT/alumina composite is investigated through TEM observations and single nanotube pullout tests. Achieving tougher ceramic composites with MWCNTs is discussed based on these results.

3.1. Pullout experiment sample preparation

The MWCNT failure during crack opening motivated our research of the crack bridging characteristics through the single nanotube pullout tests. The single nanotube pullout experiments were carried out using an *in-situ* SEM (Quanta 600 FEG; FEI) method with a nanomanipulator system (Yu et al., 2000; Yamamoto et al., 2010). An atomic force microscope (AFM) cantilever (PPP-ZEILR, nominal force constant 1.6 N/m; NANOSENSORS) was mounted at the end of a piezoelectric bender (ceramic plate bender CMBP01; Noliac) on an X–Y linear motion stage, and the composite with fracture surface (that was coated with platinum) was mounted on an opposing Z linear motion stage. The piezoelectric bender was used to measure the resonant frequency of each cantilever in vacuum. A single MWCNT on the fracture surface was clamped onto a cantilever tip by local electron-beam-induced deposition (EBID) of a carbonaceous material (Ding et al., 2005). As a precursor source for the EBID, we used n-docosane ($C_{22}H_{46}$, Alfa Aesar), which was dissolved in toluene to make a 3 mass% solution. A small amount of the solution was dropped on a cut-in-half copper TEM grid. After the solution evaporated, the TEM grid with paraffin source was mounted on the AFM chip, as shown in Fig. 9. The deposition rate of the EBID depends on several factors (Ding et al., 2005). Thus, the amount of the paraffin source, deposition time, and distance between the paraffin source and the cantilever tip were experimentally-optimized. The cantilevers serve as force-sensing elements and the spring constants of each were calculated *in-situ* prior to the pullout test using the resonance method (Sader et al., 1999). In brief, for the case of a rectangular cantilever, the force constant (k) is given by following equation,

$$k = M_e \rho_c bhL\omega_{vac}^2 \tag{5}$$

where ω_{vac} is the fundamental radial resonant frequency of the cantilever in vacuum, h, b, and L are the thickness, width, and length of the cantilever, respectively, ρ_c is the density of the cantilever (= 2.33 Mg/m³), and M_e is the normalized effective mass which takes the value $M_e = 0.2427$ for $L/b > 5$ (Sader et al., 1995). We measured ω_{vac}, h, b and L of each cantilever in the SEM and used the measured, not the nominal provided, values to calculate k. The h, b and L are determined by counting the number of pixels in the acquired SEM images. The applied force is calculated from the angle of deflection at the cantilever tip in the acquired SEM images (Ding et al., 2006). The deflection (δ) and angle of deflection (θ) at the cantilever tip are given by

$$\delta = PL^3 / 3EI \tag{6}$$

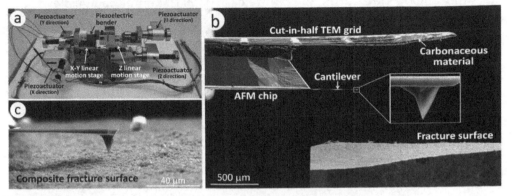

Figure 9. SEM image showing the experimental setup for pullout experiments.

$$\theta = PL^2 / 2EI \tag{7}$$

where P is the load applied at the cantilever tip, L is the cantilever length, E is the elastic modulus and I is the moment of inertia of the cantilever (Ding et al., 2006). Thus, the deflection at the cantilever tip can be represented by the angle of deflection with the following relationship (Ding et al., 2006):

$$\delta = 2\theta L / 3 \tag{8}$$

A crosshead speed – i.e., movement rate of the cantilever – of about 100 nm/s was applied for the pullout tests.

We fractured a composite specimen by conducting the fracture tests, which caused single MWCNT to project from the crack plane, as exemplified in Fig. 6a. This allows single MWCNT "pickup" with cantilever tip for subsequent tensile loading using the nanomanipulator. As mentioned above, however, the MWCNTs crossing the crack planes were strained during crack opening and possibly underwent failure, as shown in Figs. 6b, 6c and 7. Therefore, by observing the fracture surface on the composites, MWCNTs with no apparent damages were selected for the pullout tests. The physical and mechanical properties, and electrical conductivity of the composite used for the pullout testes are shown in Table 2.

Relative density (%)	Grain size (μm)	Bending strength (MPa)	Fracture toughness (MPa·m$^{1/2}$)	Hardness (GPa)	Young's modulus (GPa)	Poisson's ratio
98.9	1.43±0.31	543.8±60.9	4.74±0.12	17.0 ± 0.4	358.0	0.20

Table 2. The properties of the composite with 0.9 vol.% pristine MWCNTs. The Young's modulus and Poisson's ratio were measured by the ultrasonic pulse echo method.

3.2. Nanotube fracture during the failure of MWCNT/alumina composites

Results obtained from the pullout experiments revealed that strong load transfer was demonstrated, and no pullout behavior was observed for all 15 MWCNTs tested in this present research. Eight of these MWCNTs fractured at the composite surface and the remaining 7 MWCNTs underwent failure in the region between the fixed point on the cantilever and the crack plane, as illustrated in Fig. 10.

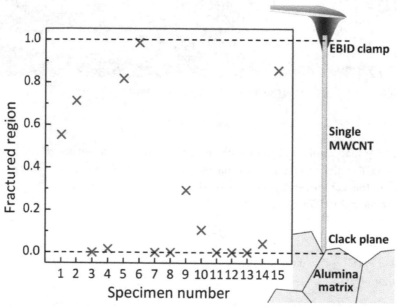

Figure 10. Fracture location of single MWCNTs under pullout loading. Of the 15 MWCNTs tested here, 8 MWCNTs fractured on the composite surface (sample numbers: 3, 4, 7, 8, 11–14) and remaining 7 MWCNTs fractured in the middle (sample numbers: 1, 2, 5, 6, 9, 10, and 15).

Two series of SEM and TEM images for each of two individual MWCNTs, captured before and after their breaking, are shown in Figs. 11 and 12. In the first series (Fig. 11; sample number 14), a MWCNT projecting 5.72 ± 0.01 μm from the fracture surface (Fig. 11a) was "welded" to a cantilever tip by local EBID, and then loaded in increments until failure. The resulting fragment attached on the cantilever tip was at least 10.9 μm long (Fig. 11b), whereas the other fragment remained lodged in a grain boundary of the alumina matrix (Fig. 11c), suggesting that MWCNT underwent failure in a sword-in-sheath manner. TEM images show a change in diameter at the location where the MWCNT underwent failure, and that the inner core protruding from the outer shells has a multi-walled closed-end structure, as shown in Figs. 11d and 11e, respectively. Given that uniformity of the interwall spacing of 0.34-nm-thick cylinder structure, approximately 11 shells underwent failure. There results strongly suggest that the MWCNTs broke in the outer shells and the inner core was then completely pulled away, leaving the companion fragment of the outer shells in the matrix. The sword-in-sheath failure did not always occur. Instead a few MWCNT failed leaving either a very short sword-in-sheath failure or a clean break. As for one example (Fig. 12; sample number 10), a MWCNT projecting 5.34 ± 0.01 µm from the crack plane (Fig. 12a) underwent failure on the composite fracture surface. The resulting fragment attached on the cantilever tip was at least 5.7 µm long (Fig. 12b), and no fragment was observed at the original position on the crack plane, suggesting that in this case the MWCNT failed by breaking inside the matrix, and did not pull out. Fig. 12c shows the TEM image of the tip of the same MWCNT which underwent very short sword-in-sheath failure or clean break during crack opening.

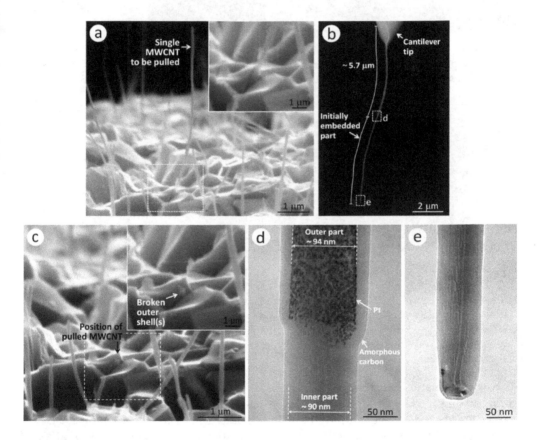

Figure 11. SEM images show (a) a free-standing MWCNT having a 5.72 ± 0.01 µm-long on the fracture surface of the composite. (b) After breaking, one fragment of the same MWCNT attached on the cantilever tip had a length ~10.9 µm. (c) The other fragment remained in the matrix. (d,e) TEM images show a change in diameter at the location where the MWCNT underwent multi-wall failure, and that it clearly has a multi-walled closed-end structure.

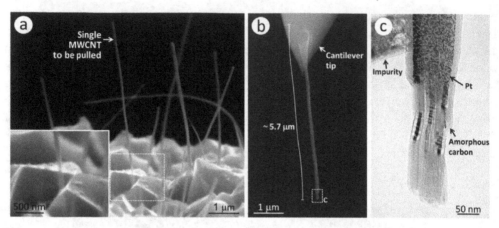

Figure 12. In the second series, (a) a tensile-loaded MWCNT with a length of 4.46 ± 0.01 μm fractured on the crack plane. (b) The resulting fragment on the cantilever tip had a length ~5.7 μm. (c) TEM image shows the MWCNT which underwent the very short sword-in-sheath failure or clean break.

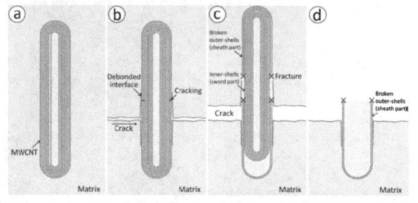

Figure 13. Schematic description of possible fracture mechanisms of the MWCNT (sample number 14). (a) Initial state of a MWCNT. (b) Tensile stresses lead to matrix crack and partial debonding formation. (c,d) As displacement increases, the MWCNTs, rather than pulling out from the alumina matrix, undergo failure in the outer shells and the inner core is pulled away, leaving the fragment of the outer shells in the matrix.

Next, we schematically describe possible processes and mechanics, explaining the MWCNT failure during crack opening (Fig. 13). As for one example, considering the sample number 14 (Fig. 11), the initial state of the MWCNT in an ideal case is a completely impregnated and isolated embedded in the matrix (Fig. 13a). Tensile stresses parallel to the axis of MWCNT length lead to matrix crack formation. Subsequently, interfacial debonding between two phases may occur (Fig. 13b), perhaps over a limited distance (but this is unlikely to make a major contribution to the fracture energy.). Since there is variability in the MWCNT strength in the debonded region on either side of the crack plane, and it is possible for the MWCNT to break at a certain position, when the stress in the MWCNT reaches a critical value. As displacement increases, the MWCNTs, rather than pulling out from the alumina matrix, undergo failure in the outer shells and the inner core is pulled away, leaving the fragments of the outer shells in the matrix (Figs. 13c and 13d).

4. Conclusion

Creating tough, fracture-resistant ceramics has been a central focus of MWCNT/ceramic composites research. In this research, the MWCNT/alumina composite with enhanced mechanical properties of 689.6 ± 29.1 MPa for bending strength and 5.90 ± 0.27 MPa·m$^{1/2}$ for fracture toughness have been successfully prepared by a novel processing method. A combination of the precursor method for synthesis of the alumina matrix, the acid treatment of the pristine MWCNTs and the spark plasma sintering method can diminish the phase segregation of MWCNTs, and render MWCNT/alumina composites highly homogeneous. The universality of the method developed here will be applicable to a wide range of functional materials such as tribomaterials, electromagnetic wave absorption materials, electrostrictive materials, and so on. Our present work may give a promising future for the application of MWCNTs in reinforcing structural ceramic components and other materials systems such as polymer- and metal-based composites.

We have also shown from TEM observations and single nanotube pullout experiments on the MWCNT/alumina composites that strong load transfer was revealed, and no MWCNT pullout behavior was observed. It is well recognized the fracture properties of fiber-reinforced composites are dominated by the fiber bridging force resulting from debonding and sliding resistance, which dictates the major contribution to the strength and toughness (Evans, 1990; Hull & Clyne, 1996). The results reported here suggest that modest improvements in toughness reported previously may be due to the way MWCNT's fail during crack opening in the MWCNT/ceramic composites. Our finding suggests important implications for the design of tougher ceramic composites with MWCNTs. The important factor for such tougher ceramic composites will thus be the use of MWCNT having a much higher load carrying capacity (as well as a good dispersion in the matrix).

Author details

Go Yamamoto and Toshiyuki Hashida
Fracture and Reliability Research Institute (FRRI), Tohoku University, Japan

Acknowledgement

The authors thank our colleague, Dr. M. Omori, Mr. K. Shirasu, Mr. Y. Nozaka, Mr. Y. Aizawa and Ms. N. Suzuki of Fracture and Reliability Research Institute (FRRI), Tohoku University, for their helpful discussions, and Mr. T. Miyazaki of Technical Division, School of Engineering, Tohoku University, for technical assistance in the TEM analysis. The authors acknowledge Prof. R.S. Ruoff of The University of Texas at Austin for his useful guidance. This work is partially supported by Grand-in-Aids for Scientific Research (Nos. 23860004 and 21226004) from the Japanese Ministry of Education, Culture, Sports, Science and Technology. This work is performed under the inter-university cooperative research program of the Advanced Research Center of Metallic Glasses, Institute for materials Research, Tohoku University.

5. References

Chen, J.; Hamon, M.A.; Hu, H.; Chen, Y.; Rao, A.M.; Eklund, P.C. & Haddon, R.C. (1998). Solution properties of single-walled carbon nanotubes. *Science*, Vol. 282, pp. 95-98.

Chen, Y.L.; Liu, B.; Huang, Y.; & Hwang, K.C. (2011). Fracture toughness of carbon nanotube-reinforced metal- and ceramic-matrix composites. *Journal of Nanomaterials*, Vol. 2011, Article ID 746029.

Cho, J.; Boccaccini, A.R. & Shaffer, M.S.P. (2009). Ceramic matrix composites containing carbon nanotubes. Journal of Materials Science, Vol. 44, pp. 1934–1951.

Dai, H.J.; Wong, E.W. & Lieber, C.M. (1996). Probing electrical transport in nanomaterials: conductivity of individual carbon nanotubes. *Science*, Vol. 272, pp. 523–526.

De Andrade, M.J.; Lima, M.D.; Bergmann, C.P.; Ramminger, G.D.O.; Balzaretti, N.M.; Costa, T.M.H. & Gallas, M.R. (2008). Carbon nanotube/silica composites obtained by sol-gel and high-pressure techniques. *Nanotechnology*, Vol. 19, article number: 265607.

Ding, W.; Dikin, D.A.; Chen, X.; Piner, R.D.; Ruoff, R.S.; Zussman, E.; Wang, X. & Li, X. (2005). Mechanics of hydrogenated amorphous carbon deposits from electron-beam-induced deposition of a paraffin precursor. *Journal of Applied Physics*, Vol. 98, article number: 014905.

Ding, W.Q.; Calabri, L.; Chen, X.Q.; Kohhaas, K.M. & Ruoff, R.S. (2006). Mechanics of crystalline boron nanowires. *Composites Science and Technology*, Vol. 66, pp. 1112–1124.

Ebbesen, T.W.; Lezec, H.J.; Hiura, H.; Bennett, J.W.; Ghaemi, H.F. & Thio, T. (1996). Electrical conductivity of individual carbon nanotubes. *Nature*, Vol. 382, pp. 54–56.

Esumi, K.; Ishigami, M.; Nakajima, A.: Sawada, K. & Honda, H. (1996). Chemical treatment of carbon nanotubes. *Carbon*, Vol. 34, pp. 279-281.

Evans, A.G. (1990) Perspective on the development of high-toughness ceramics. Journal of the American Ceramic Society, Vol. 73, pp. 187–206.

Fan, J.P.; Zhao, D.Q.; Wu, M.S.; Xu, Z. & Song, J. (2006). Preparation and microstructure of multiwalled carbon nanotubes-toughened composite. *Journal of the American Ceramic Society*, Vol. 89, pp. 750–753.

Fan J.P; Zhuang, D.M.; Zhao, D.Q.; Zhang, G.; Wu, M.S.; Wei, F. & Fan, Z.J. (2006). Toughening and reinforcing alumina matrix composite with single-wall carbon nanotubes. *Applied Physics Letters*, Vol. 89, pp. 121910-1219103.

Gonzalez-Julian, J.; Miranzo, P.; Osendi, M.I. & Belmonte, M. (2011). Carbon nanotubes functionalization process for developing ceramic matrix nanocomposites. *Journal of Materials Chemistry*, Vol. 21, pp. 6063-6071.

Huang, J.Y.; Chen, S.; Wang, Z.Q.; Kempa, K.; Wang, Y.M.; Jo, S.H.; Chen, G.; Dresselhaus, M.S. & Ren, Z.F. (2006). Superplastic carbon nanotubes – conditions have been discovered that allow extensive deformation of rigid singlewalled nanotubes. *Nature*, Vol. 439, p. 281.

Hull, D & Clyne T.W. (1996). *An Introduction to Composite Materials (Second edition)*. Cambridge University Press, 0521388554, The Edinburgh Building, Cambridge CB2 2RU, UK.

Japanese Industrial Standards (JIS). (1995). R 1607.

Kita, J.; Suemasu, H.; Davies, I.J.; Koda, S. & Itatani, K. (2010). Fabrication of silicon carbide composites with carbon nanofiber addition and their fracture toughness. *Journal of Materials Science*, Vol. 45, pp. 6052-6058.

Liu, J.; Rinzler, A.G.;Dai, H.; Hafner, J.H.; Bradley, R.K.; Boul, P.J.; Lu, A.; Iverson, T.; Shelimov, K.; Huffman, C.B.; Rodriguez-Macias, F.; Shon, Y.S.; Lee, T.R.; Colbert, D.T. & Smalley, R.E. (1998). Fullerene pipes. *Science*, Vol. 280, pp. 1253-1256.

Ma, R.Z.; Wu, J.; Wei, B.Q.; Liang, J. & Wu, D.H. (1998). Processing and properties of carbon nanotubes–nano-SiC ceramic. *Journal of Materials Science*, Vol. 33, pp. 5243-5246.

Miyahara, N.; Yamaishi, K.; Mutoh, Y.; Uematsu, K. & Inoue, M. (1994). Effects of grain size on strength and fracture toughness in alumina, *JSME International journal Vol. 37*, pp. 231-237.

Mukerji, J. (1993) Ceramic matrix composites. *Defence Science Journal*, Vol. 43, pp. 385–395.

Omori, M. (2000). Sintering, consolidation, reaction and crystal growth by the spark plasma system (SPS). *Materials Science and Engineering A*, Vol. 287, pp. 183-188.

Peigney, A. (2003). Composite materials: tougher ceramics with nanotubes. *Nature Materials*, Vol. 2, pp. 15–16.

Peigney, A.; Flahaut, E.; Laurent, Ch.; Chastel, F. & Rousset, A. (2002). Aligned carbon nanotubes in ceramics-matrix nanocomposites prepared by high-temperature extrusion. *Chemical Physics Letters*, Vol. 352, pp. 20–25.

Peng, B.; Locascio, M.; Zapol, P.; Li, S.Y.; Mielke, S.L.; Schatz, G.C. & Espinosa, H.D. (2008). Measurements of near-ultimate strength for multiwalled carbon nanotubes and irradiation-induced crosslinking improvements. *Nature Nanotechnology*, Vol. 3, pp. 626–631.

Rice, R.W. (1996). Grain size and porosity dependence of ceramic fracture energy and toughness at 22°C, *Journal of Materials Science*, Vol. 31, pp. 1969-1983.

Sader, J.E.; Chon, J.W.M. & Mulvaney, P. (1999). Calibration of rectangular atomic force microscope cantilevers. *Review of Scientific Instruments*, Vol. 70, pp. 3967–3969.

Sader, J.E.; Larson, I.; Mulvaney, P. & White, L.R. (1995). Method for the calibration of atomic force microscope cantilevers. *Review of Scientific Instruments*, Vol. 66, pp. 3789-3798.

Sheldon, B.W. & Curtin, W.A. (2004). Nanoceramic composites: tough to test. *Nature Materials*, Vol. 3, pp. 505–506.

Sun, L.; Gao, L. & Li, X. (2002). Colloidal processing of carbon nanotube/alumina composites. *Chemistry of Materials*, Vol. 14, pp. 5169–5172.

Sun, J.; Gao, L. & Xihai Jin, X. (2005).Reinforcement of alumina matrix with multi-walled carbon nanotubes. *Ceramics International*, Vol. 31, pp. 893–896.

Treacy, M.M.J.; Ebbesen, T.W. & Gibson, J.M. (1996). Exceptionally high Young's modulus observed for individual carbon nanotubes. *Nature*, Vol. 381, pp. 678–680.

Wang, X.; Padture, N.P. & Tanaka, H. (2004). Contact-damage-resistant ceramic/single-wall carbon nanotubes and ceramic/graphite composites. *Nature Materials*, Vol. 3, pp. 539-544.

Yamamoto, G.; Omori, M.; Yokomizo, K.; Hashida, T. & Adachi, K. (2008). Structural characterization and frictional properties of carbon nanotube/alumina composites prepared by precursor method. *Materials Science and Engineering B*, Vol. 148, pp. 265–269.

Yamamoto, G.; Omori, M.; Hashida, T. & Kimura, H. (2008). A novel structure for carbon nanotube reinforced alumina composites with improved mechanical properties. *Nanotechnology*, Vol. 19, article number: 315708.

Yamamoto, G.; Omori, M.; Yokomizo, K. & Hashida, T. (2008). Mechanical properties and structural characterization of carbon nanotube/alumina composites prepared by precursor method. *Diamond and Related Materials*, Vol. 17, pp. 1554-1557.

Yamamoto, G.; Suk, J.W.; An, J.; Piner, R.D.; Hashida, T.; Takagi, T. & Ruoff, R.S. (2010). The influence of nanoscale defects on the fracture of multi-walled carbon nanotubes under tensile loading. *Diamond and Related Materials*, Vol. 19, pp. 748-751.

Yamamoto, G.; Shirasu, K.; Hashida, T.; Takagi, T.; Suk, J.W.; An, J.; Piner, R.D. & Ruoff, R.S. (2011). Nanotube fracture during the failure of carbon nanotube/alumina composites. *Carbon*, Vol. 49, pp. 3709-3716.

Yao, W.; Liu, J.; Holland, T.B.; Huang, L.; Xiong, Y.; Schoenung, J.M. & Mukherjee, A.K. (2011). Grain size dependence of fracture toughness for fine grained alumina, *Scripta Materialia*, Vol. 65, pp. 143–146.

Yu, M.F.; Lourie, O.; Dyer, M.J.; Moloni, K.; Kelly, T.F. & Ruoff, R.S. (2000). Strength and breaking mechanism of multiwalled carbon nanotubes under tensile load. *Science*, Vol. 287, pp. 637-640.

Zhan, G.D.; Kuntz, J.D.; Wan, J. & Mukherjee, A.K. (2003). Single-wall carbon nanotubes as attractive toughening agents in alumina-based nanocomposites. *Nature Materials*, Vol. 2, pp. 38-42.

Chapter 10

Characterisation of Aluminium Matrix Syntactic Foams Under Static and Dynamic Loading

M. Altenaiji, G.K. Schleyer and Y.Y. Zhao

Additional information is available at the end of the chapter

1. Introduction

The resistance of engineering structures, subjected to blast and impact loads, is of great interest within the engineering community and government agencies. The interest and importance of such material stem from the need for providing protection against possible terrorist threats. Development of a light -weight, strong and ductile material capable of providing protection of vehicles and occupants against impact and blast, however, is a formidable challenge facing the materials community.

In a blast or an impact, a structure usually undergoes large plastic deformation or break. The important characteristics of structural response include: (i) mode of deformation and failure, (ii) impulse transfer, and (iii) energy absorption in plastic deformation (Hanssen et al., 2002). To characterize cellular materials, the characteristics of the base material of the foam, its relative density, type of its cells (open or closed) and the mean cell diameter, all of these must be known (Ashby et al, 1997). The characterization and testing of metallic foams require special precautions. A cellular material could be characterized by several parameters such as: the constituent raw materials, the mean cell diameter, relative density (porosity), and cell size and shape etc. The constituents can be analysed by using x-ray tomography, optical microscopy and scanning electron microscopy. The easiest way to measure porosity (relative density) is to weigh the sample of a known volume. Additionally, metal foam could be characterized by its cell topology (open cells, closed-cells). Characterizations of metal foam using optical microscopy can be carried out provide that the foam is fully impregnated with opaque epoxy resin before polishing. Scanning electron microscopy has been found to be most informative for open cells foams rather than for the closed-cell foams (Ashby et al., 2000). On the other hand, x-ray tomography technique is a good way to investigate the deformation modes of cellular solids. It depends on the low overall absorption of x-rays as such low absorption allows large specimens to be cut into small pieces. The second advantage of this technique is the fact that the deformation could be monitored non-destructively (Fazekas et. al, 2003).

The mechanical properties of cellular materials can be characterized by the compression test. The behaviour of the cellular material under static- compression is known to be completely different from that of the solid material without any compression. This test is also used for the study of the behaviour of material under crushing load. The compression test is useful for measuring plastic flow, ductile fracture limit and compressive fracture properties of brittle materials. In a compression, several modes of deformation can may occur as illustrated in Figure.1. These are as follows: a) buckling mode when the ratio of the sample height (h) to its width (w); i.e. (h/w) > 5; b) shearing mode may appear when (h/w) ~ 2.5; and, c) double barrelling when ratio of length (L) and diameter (D) i.e. (L/D) > 2 and friction is presented at contact surfaces, d) barrelling mode when (h/w) < 2; e) a homogenous compression mode when the ratio (h/w) lies between 2.0 and 1.5 and finally; f) the possibility of compressive instability (Medlin & Kuhn, 2000).

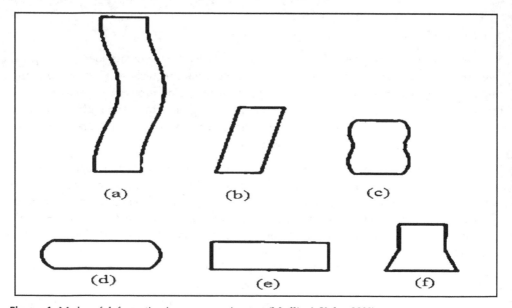

Figure 1. Modes of deformation in a compression test (Medlin & Kuhn, 2000).

The study of the compressive behaviour of cellular materials offers the benefit of knowledge on most of the mechanical properties of the material. The mechanical properties of cellular materials are different to its base material due to the difference in structures. The following section explains the compressive properties of cellular materials and its relevance to their myriad applications.

Stress-strain curve is used to determine Young's modulus which, in turn, yields values of strength and strain densification. An example of a typical stress-strain curve (Gibson, 2000) of cellular material under uni-axial compression test condition is shown in Figure.2. Stress-strain curve of metallic foam shows three regimes of material behaviour. As can be seen in the graph, initially it is a linear dependence of elasticity (strain) with stress. Such dependency is governed by the strength of cell walls of the base material. As is expected, the stiffness of the metallic foam will be higher with the increase of strength of the cell wall. It has been observed that the type and grade of metallic alloy as the base material dictate the stiffness of the foam structure. It has been found that the metallic alloy which has higher yield strength, gives higher stiffness than other alloys. At this stage, the material undergoes a non-permanent deformation. Usually, the deformation is in the form of cell face stretching and consequently extending and stretching the cell edges. This, in turn, increases the stiffness which was driven by the cells edges in closed-cell foam configuration while extending and bending deformation mode in open cell foam configuration.

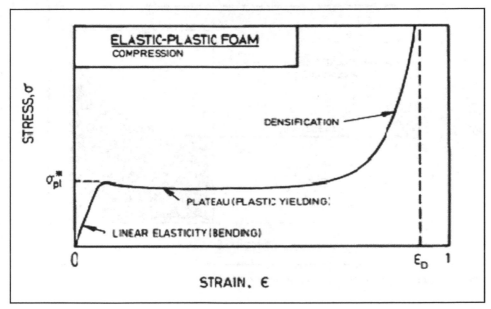

Figure 2. A typical stress-strain curve for foam showing linear elastic, stress plateau and densification regimes (Gibson, 2000).

Energy absorption is the ability of the material to convert the kinetic energy into energy of some other form such as, heat, viscosity, visco-elasticity, friction etc. The kinetic energy must be less than the maximum limit of energy absorption of the material to keep the object safe. Also, the energy absorption property must be multi-directional, i.e. allowing absorption of the impact from any direction. Cellular materials have better capacity of energy absorption than its base materials. Foams have a capability to absorb the kinetic energy by bending, buckling or fracture (plastically) of the cell walls depending on the characteristics of the base material of the foam (Ashby, 1997).

Energy absorption capability of foams depends on the stress and strain of the plateau in stress-strain curve. Foams have long, flat stress-strain curve where the cell walls collapses plastically to nominal strain ε_D at a constant stress and the phenomenon is called, 'the plateau stress up'. The plateau stress must be below the value that causes damage to the object. The best energy absorbing material is that which has the longest plateau and absorbs the most energy before reaching the densification strain. The material that has a long and flat stress-strain curve is considered to be an ideal energy absorber. The energy absorber capability of the material is measured by the length and height of the flat stress-strain curve. The area under the plateau of the stress-strain curve represents energy per unit volume which can be absorbed as illustrated in Figure 3. The area under the flat part of the curves is the useful energy per unit volume, W_v that can be absorbed.

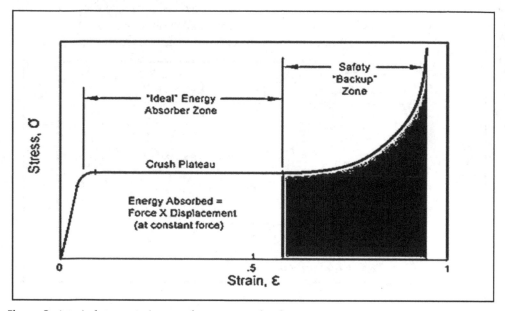

Figure 3. A typical stress-strain curve for an energy absorber.

2. Metal matrix syntactic foam

Matrix syntactic foams are composite materials consisting of a matrix implanted with hollow or porous ceramic particles. These foams are a new class of materials which are manufactured by a variety of metal or polymer matrices and micro-spheres ceramics. It consists of a metal matrix and micro-spheres ceramics, embedded in the matrix. A metal matrix could be of aluminium, steel, titanium or magnesium. Ceramic micro-sphere could be a porous or hollow structure, but hollow metal spheres are rarely used.

The size of the hollow ceramics micro-sphere determines the porosity of the matrix syntactic foam. In addition, it has an effect on the strength of the syntactic foam. Therefore, the matrix syntactic foam has different physical and mechanical properties than other cellular materials. Matrix syntactic foam is considered as a light material having high energy absorption capacity. It has been used in automotive, naval, aerospace and other industries where 'light-weight' property is an essential requirement. In addition to that it could be used for reducing shock loadings associated with mine blast on military vehicles where the high energy absorption capability and light weight property are needed.

2.1. Fabrication process

Metal matrix syntactic foams consist of a combination of ceramic micro-sphere and metal matrix. Most metal matrices should be light metals like aluminium, magnesium, titanium etc. Two types of micro-sphere in common use have either porous or hollow structures. Four types of cellular spheres have so far been used to fabricate metal matrix syntactic foam as follows: (i). amorphous silica, (ii). Al2O3 spheres, (iii). CMs of crystalline mullite and (vi). steel spheres (Tao, 2010).

There are two main ways to fabricate metal matrix syntactic foams as follows: i) stir casting with spray processes, and ii) infiltration casting using liquid state. In stir casting, metal matrix syntactic foams are fabricated by mixing a liquid of metal matrix with the ceramic particles followed by casting. This method is very simple and cheap, but inhomogeneous structure of syntactic foam has been found to appear due to float of ceramic spheres to the top of the melt. In infiltration casting, the metal matrix is placed above the ceramic spheres and is pressed to infiltrate into the ceramic sphere where it is solidified to produce metal matrix syntactic foam. The infiltration casting can be conducted by gas pressure or die casting. This method has the advantage that the matrix and ceramic spheres are well bonded and the micro-spheres are usually uniformly distributed.

2.2. Porosity

There are two classifications for porosity: open or closed types. The open porosity is defined by the ratio of the volume of void space that is accessible from exterior to bulk volume. Meanwhile, the closed porosity is defined by the ratio of the volume of void space that is not accessible from exterior to bulk volume. The porosity of foams depends on the shape and size of the pores. The porosity of metallic syntactic foams, however, is determined by the porosity of the cellular spheres. The strength of the micro-sphere of syntactic foam could be tailored through the appropriate selection of the wall thickness and the radius of micro-spheres (Kiser, 1999). The porosity of cellular materials controls the plateau strength and energy absorption capacity of the foams. In general, the strength of porous materials increases with the reduction of porosity.

The main parameter that has the most significant effect on the porosity of the metal matrix syntactic foam is the porosity of the ceramics micro-spheres. The radius, shell thickness and the volume fraction of the hollow spheres are the parameters that control the porosity of hollow spheres. The porosity of a ceramic sphere could be estimated by the following equation (Kiser et. al, 1999):

$$\theta_s = (1 - \frac{\rho_s}{\rho_o}) \tag{1}$$

Where, θ_s is the porosity of ceramic material, ρ_s is its effective density and ρ_0 is the density of the solid part of the spheres. In principle the porosity of metal matrix syntactic foam may be calculated using the formula below (Zhang& Zhao, 2007):

$$\theta_f = (1 - f_{al})\theta_s \tag{2}$$

Where, θ_f is the porosity of syntactic foam and f_{al} is the volume fraction of metal matrix. The authors have developed a general formula that has been widely used to calculate the porosity of metallic syntactic foam for all types of spheres and is given below:

$$\theta_s = \frac{\rho_m - \rho_f}{\rho_m \rho_s}\left(1 - \frac{\rho_s}{\rho_o}\right) \tag{3}$$

Where, ρ_m, ρ_f are ρ_0 are the density of metal matrix, syntactic foam and solid part of cellular spheres respectively, and ρ_s is the effective density of the cellular spheres (Tao, 2010). It was reported earlier (Kiser et. al, 1999) that the thickness and the radius of the shell of the hollow sphere and the micro-balloon do control the composite porosity ρ_0 and developed a formula to calculate the porosity of syntactic foams with hollow spheres as follows:

$$\rho_o = f(1 - \frac{t}{R})^3 \tag{4}$$

Where, ρ_0 is the porosity, f is the volume percentage of the hollow spheres, and t is the shell thickness and, R is the radius of the hollow spheres. It was also reported [Kiser et.al, 1999] that the strength of the micro-balloon could be tailored through appropriate selection of the parameter, [t/R], having the strength increasing with increasing factor, [t/R].

2.3. Compressive strength of metallic matrix syntactic foam

The compressive strength of metallic matrix syntactic foam is controlled by the strength of metal matrix and the ceramic particles. The parameters such as: volume fraction, structure and distribution of the ceramics particles have considerable effect on the properties of syntactic foams (Zhao& Tao, 2009). It has been reported (Rohatgi et. al, 2006) that the compressive strength of the metallic syntactic foams decreases with the increase in volume fraction of the micro-spheres ceramics. In fact, the compressive strength of metal matrix has been found to be higher than that of the micro-spheres ceramics. Consequently the strength of the metallic matrix syntactic foam decreases with the increase in the volume fraction of ceramics particles.

It has been reported that the yield strength of metallic foam decreases with the decrease of the size of micro-spheres ceramic (Rohatgi et. al, 2006). Moreover, it has been noted (Altenaiji et.al, 2011) that the smaller the size of the micro-sphere ceramic, the weaker is the strength of metal matrix syntactic foam. There are also other parameters, such as the strength of the metal matrix, which effect the strength of metal matrix syntactic foam as illustrated in Figure 4 and Figure 5.

In contrast, Palmer reported (Palmer et. al, 2007) that, the lower compressive strength of metallic syntactic foams is associated with larger micro-spheres. It was however predicted

(Zhao& Tao, 2009) that there should be a difference in the compressive strength of different void contents in different sized ceramic spheres. It was also predicted (Sun et.al, 2007) that the thickness and the radius of micro-spheres ceramics have an effect on the compressive strength of metallic syntactic foams. The micro-sphere ceramics which have higher wall thickness to radius ratios (t/r) were found to have higher compressive strength of the resultant. Earlier reports (Kiser et. al, 1999) indicated that the compressive strength of metal matrix syntactic foam increased from 70 to 230 MPa with the increase of the factor, (t/r) from 0.12 to 0.48, i.e. an increase by a factor of 3 for a similar factor of increase in the pressure.

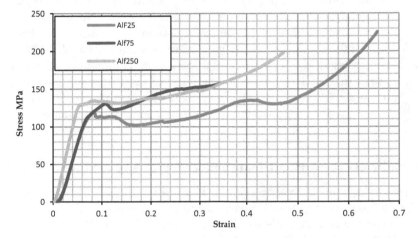

Figure 4. Comparison of stress-strain curve of aluminium syntactic foam with different ceramic micro-sphere size.

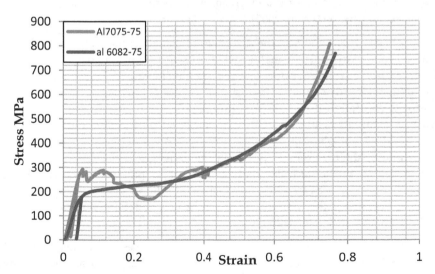

Figure 5. Comparison of stress-strain curve of aluminium syntactic foam with different aluminium matrix.

The compressive strength of metallic matrix foams is also affected by the types of micro-spheres ceramics. It has been found that metallic matrix foam containing hollow ceramics spheres have much higher compressive strength than those containing porous ceramics spheres (Tao et.al, 2009). The compressive behaviour of metal matrix syntactic foams with hollow steel spheres of same thickness but different composition was studied (Vendra & Rabiei, 2007). It was observed that the metal matrix, containing low carbon steel spheres had much lower compressive strength than the ones which had stainless steel spheres.

It has been reported that the compressive strength of the metal matrix syntactic foams is also affected by the type of metal matrix. Studies on the compressive behaviour of syntactic foams with same ceramics micro-sphere but different aluminium matrix had been conducted. It was found that the compressive strength of al7075-t is double that of al 6082-T with same method of processing. It was also indicated that when the aluminium matrix was replaced by another metal matrix the compressive strength of metallic matrix syntactic foams changed. When the al 7075 was used the compressive strength of metallic matrix syntactic foams had increased to approximately double than that containing pure aluminium (Balch& Dunand, 2006).

2.4. Failure of metallic matrix syntactic foam under compressive loading

As is illustrated in Figure.2, metallic matrix syntactic foam behaved like any foam under compression loading. It has three regimes of behaviour under compression loading. Initially it starts with a linear elasticity and then follows by plateau strength and finally, ends by the densification strain. But, the failure behaviour of metal matrix syntactic foams is different due to it compositions. Its failure is controlled by different plastic characteristics. It can be either ductile, susceptible to collapse under crushing of micro-spheres ceramics or brittle, susceptible to shear failure (Balch et.al, 2005).

It was reported (Zhao& Tao, 2009) that the three factors that affect the failure behaviour of metal matrix foam are: i) ductility of metal matrix, ii) structure of micro-spheres ceramic and, iii) volume fraction of micro-spheres ceramic and metal matrix. It is also reported that the behaviour of 'metal matrix syntactic' foams' performance failure depends on the type of the metal matrix. It was found that the aluminium alloy-T6 had two localized deformation bands of crushed materials while, the CP-Al (commercial purity aluminium) exhibited transition into densification plateau, showing extensive barrelling and pore deformation over a large volume of the material. Additionally, the damages of these two foams were reported to be different. In the alloy-T6, damage was concentrated in two much thicker crush band where the strain was very high. In contrast, in the CP-Al foams' damage was spread over a big volume of the material, resulting in plastic deformation of the matrix along with sphere fracture (Balch et.al, 2005).

It has been reported (Tao & Zhao, 2009) that the compressive failure of metallic matrix syntactic foam is affected by the volume fraction of metal matrix micro-spheres ceramic. High volume of metal matrix causes ductile failure in the form of collapse of the material. In contrast, low volume of the metal matrix tends to cause brittle failure in the form of shear.

As mentioned previously, the thickness and the radius of micro-spheres ceramics have an effect on the compressive strength of metallic syntactic foams. The micro-sphere ceramics with higher wall thickness-to-radius ratios, (t/r) were found to have higher compressive strength for the resultant (Sun et.al, 2007). The syntactic foam metal matrix with low values of the ratio, (t/r) failed as brittle failure form while those with higher values failed as ductile failure form also known as crushing or collapse failure form (Kiser et.al,1999) . In contrast, another report (Sun et.al, 2007) stipulated that the metal matrix syntactic foam with low values of the ratio, (t/r), failed as ductile in the form of crushing or collapse failure while that had higher values failed as brittle in the form of shear failure. It was, however, reported earlier (Gupta et.al, 2004) that different types of syntactic foams in which the matrix materials and the radius of micro-spheres were fixed while the wall thickness of the micro-spheres was varied. It had been found that failure type of different matrix syntactic foams was similar. In general, therefore, it may be concluded from the present up-to-date literature review that the value of the wall thickness to radius ratio (t/r) dictates the form of the compressive failure of metal matrix syntactic foams while the value of the ratio is dependent upon the factor that affect the compressive failure. Furthermore, the strength of the micro-spheres ceramic has an effect on the strength of metal matrix syntactic foams.

3. Dynamic compressive testing of matrix syntactic foam

Metal matrix syntactic foams have been compressive tested under dynamic loading. Such loadings can be carried out through, either low or high speed impact. It has been studied under drop weight and split pressure Hopkinson bar test methods. It has been observed (Dou et.al, 2007) that the dynamic stress-strain curve has three regimes as the quasi-static stress strain curve. Meanwhile, the yield strength of metal matrix syntactic foams under dynamic loading is ~ 45-60% higher than that of quasi static compression. Moreover, Zhang investigated (Zhang& Zhao,2007) four samples of aluminium matrix syntactic foams under impact and reported that many oscillations appeared at the beginning of stress strain curve where the strain is low due to the high vertical vibration of the drop hammer. In addition, it was found that, out of four samples tested two had 10%-30% higher plateau stress than that of a quasi-static compression but the other two had lower plateau stress than that of the quasi-static compression. Therefore, the plateau strength is mostly determined by the volume fraction of metal matrix in the metallic syntactic foam. It was evaluated from the behaviour of aluminium matrix syntactic foams under high impact loading (Balch et al., 2005). Split pressure Hopkinson bar technique was used to characterize the material under dynamic loading. It was concluded that, the behaviour of aluminium matrix syntactic foams at high strain had higher peak strength and plateau stress than those measured during quasi-static testing. In addition, the peak strength of the stress strain curve has been found to have shifted slightly at higher strain value.

3.1. Low velocity dynamic loading

Low velocity dynamic loading has been conducted by using many facilities that include the Charpy and Izod pendulums, the drop weight fixture such as the Gardner as well as hydraulic machines, designed to conduct both in plane and out of plane testing at velocities up to 10 m s⁻¹(Cantwell& Morton, 1991) The Charpy pendulum is simple to use and its design scheme is presented in Figure. 6. Usually, thick beam specimen is used which has a notch in the middle. The notch must be fixed opposite to the impact of the swing pendulum. The energy dissipated during impact is calculated by multiplication of the weight of the pendulum and the difference of the pendulum height in each side. The disadvantage of this test is that the high frequency oscillation on 'load versus time' curves due to natural frequency of the hammer does introduce error.

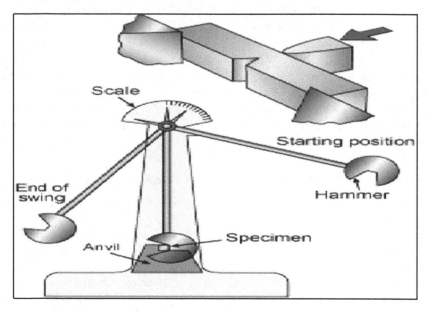

Figure 6. Schematic of the Charpy pendulum-type tester.

Comparatively, the Izod test is similar to the Charpy pendulum test except that the notch is located at the end of the specimen and the hammer impacts at the free end (Ellis, 1996). Cantwell and Morton suggested that Chapy and Izod tests both are suitable for evaluating the impact performance of the materials and a step in determining the dynamic toughness of the materials. In addition, drop-weight impact test have been used to conduct low velocity dynamics compressive testing of composites at strain rates ranging from 10 to several hundred per seconds (Hsiao et.al, 1998a, 1998b).

The heavy weight is guided by two smooth steel rails and fall from a height to strike the specimen (Fanjing, 2010). The impact of falling weight does not cause destruction of the test specimen but rebounds where the residual energy could be determined (Cantwell& Morton, 1991) as shown in Figure.7.

A piezoelectric load cell is used to measure the variation of load with time during impact. It is located underneath the base on the test rig (Yiou, 2009). While, the incident velocity of the impactor can be determined from the equation of motion as follows:

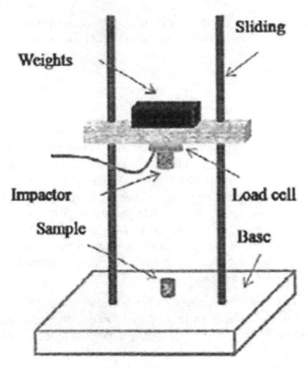

Figure 7. Schematic of the drop- hammer weight rig.

$$v=\sqrt{(2\,g\,h)} \tag{5}$$

$$\frac{a}{g} = \frac{v\omega}{gD} \tag{6}$$

where V, w and D are the weight velocity, natural frequency and dynamic load factor (usually equal to 1.77) respectively (Rajendran et.al, 2008).

And predicted deflection is given as:

$$\delta = \frac{mgh}{\sigma_{pl}\,A} \tag{7}$$

where, m, h, g, σ_{pl} and A are the weight mass, height of drop weight, gravitational force ,plateau stress and cross section area of sample respectively. The advantage of the drop-weight impact test with respect to Chapy and Izod test is that a wider range of sample geometries can be tested by changing the support size and shape. However, the stress wave reflection effect on the stress-time response of the specimen is observed in this method. Also, the limitation of strain rate which depends on the speed of falling weight was found to be directly linked to the drop height (Barre et.al, 1996). Furthermore, hydraulic test machines have been used to characterise the deformation and failure of the materials at high strain rate (Beguefin & Barbezat, 1989). One advantage of this type of testing machine is that it practically removes the problem of vibration noise that appears when using the drop weight test. Additionally, the limitation of the strain rate is extended up to a value of 50 s^{-1} from the static strain rate value. However, caution should be exercised to ensure that mass of the load cell is as low as possible to avoid the concealment of the true material response (Beguefin & Barbezat, 1989).

3.2. Medium and high velocity dynamics loading

Medium and high velocity dynamics loading has been conducted by using many facilities that include Hopkinson bar and Gas gun. In the Hopkinson bar technique, the induced wave propagation in a long elastic metallic bar has been used to measure the pressure produced during dynamic events (Kolsky, 1949). The Hopkinson bar test was used to measure the dynamic stress-strain response of materials. A high precision strain gages, signal conditioners, and high speed digital oscilloscopes are used in Hopkinson bar with high sensitivity and accuracy. The sensitivity of the pressure bar is determined by the properties of the material of the bar such as the density and the elastic wave impedance (Ramesh, 2008). In addition, the type of stain gages and the characteristics of the associated instrumentations have an effect on the sensitivity.

To determine the properties of the material under dynamics loading the procedures for the Hopkinson bar technique is similar to that of the dynamic technique mentioned earlier. These are used to provide complete stress strain data as a function of strain rate. This technique has been used to determine the dynamic properties of numerous engineering materials like ferrous and non-ferrous alloys, polymers, ceramics, and concrete (Tan et.al, 2005). Additionally, it has been used to characterise soft materials and metal foams that have low mechanical impedance, required for increasing the sensitivity of testing device (Dung et.al, 2011; Gay III et.al, 2000; Chen et.al, 1997). Split Hopkinson pressure bar has been used to measure the compressive mechanical behaviour of materials, to load samples with uni-axial tension (Klepaczco et.al, 1997) and to measure compression torsion (Gilchrist, 2009). Also, it has been designed to measure the fracture toughness of an impact loaded material (Krauthauser, 2003).

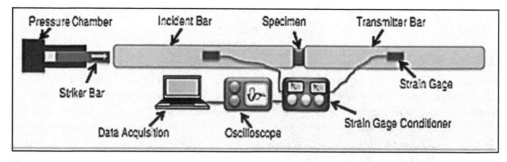

Figure 8. Schematic SHPB (Split Hopkinson pressure bar) setup for high strain rate compression testing

Before providing impact in an SHPB device the velocity of the striker was measured using an electronic velocity measurement unit. The gas gun pressure was varying that cause the variant impact velocities. The incident and transmitter bars in the device are made of steel having a length of 1000 mm and a diameter of 20 mm. The bar's mechanical properties are: Young's modulus, 207 GPa, elastic wave speed, 5890 m/s and Poisson's ratio, 0.29.

In the Hopkinson bar technique, the induced wave propagation in a long elastic metallic bar has been used to measure the pressure produced during dynamic events (Dung et.al, 2011). The Hopkinson device is used to measure the dynamic stress-strain response of materials. A high precision strain gage, signal conditioners, and high speed digital oscilloscopes are used in Hopkinson bar which provides high sensitivity and accuracy. The sensitivity of the pressure bar is governed by the properties of the material of the bar such as, the density and the elastic wave impedance (Gray III, 2000). In addition, the type of stain gages and the characteristics of the associated instrumentations have an effect on the sensitivity.

The stress at the output bar is found by converting strain gages data (volts) to the stress by using following relations (Harrigan, 2005)

$$\sigma (t) = \frac{E_b.2\epsilon v(t)}{G_g.K_g.v_i(1+v_B)} \tag{8}$$

$$\sigma (t) = \frac{A.E.\epsilon_t(t)}{Ao} \tag{9}$$

Where, $\sigma (t)$ is the stress of transmitter bar as function of time, $\epsilon v (t)$ is the strain gage voltage as function of time, G_g is the amplification factor, K_g is stain gage factor, V_I is bridge input voltage, E_b is the elastic modulus of the bar, A is cross sectional area of the bar, A_o is the cross sectional area of the sample, $\sigma (t)$ is the transmitted axial strain pulses and v_B is Poisson's ratio of the bar material. In addition, the time dependent strain rate and strain are calculated using the following equations:

$$\epsilon^{\cdot}(t) = \frac{2\ C_b\epsilon_r\ (t)}{Io} \tag{10}$$

$$\epsilon (t) = \int_0^t \epsilon^{\cdot}(t)dt \tag{11}$$

Where C_b, I_o and A_o are the sound wave velocity, length and cross sectional area of the specimen respectively, and $\epsilon (t)$ is the reflected axial strain pulses.

The Hopkinson bar technique is used to provide completer stress strain data as a function of strain rate. This is also used to determine the dynamic properties of numerous engineering materials; like ferrous and non-ferrous alloys, polymers, ceramics, and concrete (Forrestal et.al, 2002).

Additionally, it has been widely used to characterise soft materials, metal foams that have low mechanical impedance which is required increasing the sensitivity of testing device (Klepaczco, 1997; Kiernan et.al, 2009; Lopatnikov et.al, 2003; El-Nasri, 2005; Harding, 1960).

Split Hopkinson pressure bar has been used to measure the compressive mechanical behaviour of a material, loading samples in uni-axial tension, simultaneous compression torsion (Lewis & Goldsmith, 1973).Also, it has been designed to measure the fracture toughness of an impact loaded material (Taylor, 1948). Metal matrix syntactic foams are characterised by using SHPB and the experimental setup and results are shown in Table .1 and Figure.9.

	Sample ID	Weight (g)	Length (mm)	Avg. Width (mm)	Velocity applied (m/s)
Pre		12.6	16.84	21.40 ±0.1	
Post	SHPB #2	12.6	14.26	24.00 ±0.2	21.6
Pre		12.6	16.84	21.40 ± 0.1	
Post	SHPB #3	3.6	3.69	29.00 ± 1	35 ± 1
Pre		11.5	17.04	20.60 ± 0.5	
Post	SHPB #4	7	7.74	22.15 ±1	22.35 ±1
Pre		12	15.2	20.60 ± 0.5	
Post	SHPB #6	2.1	2.63	27.50 ± 1	35

Table 1. Summary of experimental results of aluminium matrix syntactic foams under dynamic loading in SHPB.

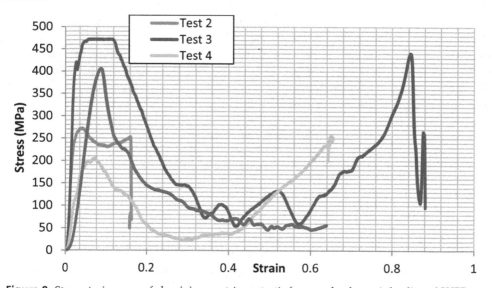

Figure 9. Stress-strain curve of aluminium matrix syntactic foam under dynamic loading of SHPB.

Dynamic compression testing showed a 10–30% increase in peak strength compared to the quasi-static results. The strain rate sensitivities of these foams are similar to those of aluminium matrix composite materials as shown in Figure.10. These foams displayed excellent energy-absorbing capability. This suggests that these have potential for use in high loading applications or for impact protection in situations where a high plateau stress is desired (Balch et.al, 2005).

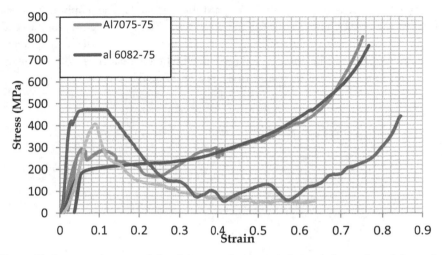

Figure 10. Stress-strain curve of aluminium matrix syntactic foam under static and dynamic loading.

High impact dynamic loading can be achieved by using high pressure gas gun (Cantwell& Morton, 1991). Gas gun apparatus contains a breech, pump tube, transition section, launch tube and target tank as illustrated in Figure 11.

In addition to high speed camera this device can also record and analyse the moment of impact and post impact behaviour (Hazell et.al, 2010) and can monitor the behaviour of the projectile material during the impact.

Figure 11. The gas gun testing apparatus (RMCS-Cranfield university)

In gas gun, a helium gas is used to accelerate a 12 mm diameter steel spheres to velocities approaching 620 m/s. The helium gas is pumped in to the breech at high pressure. An aluminium diaphragm blocks the release of the gas at the first stage. The diaphragm bursts and relieves the pressure when it reaches the required value. The helium gas is compressed which then heats up the volume of the gas by the polymer piston until the second diaphragm bursts. Consequently, the projectile and sabot is accelerated when the piston is at its final stage and the gas volume has expanded. The results of this test are shown in Figure. 12.

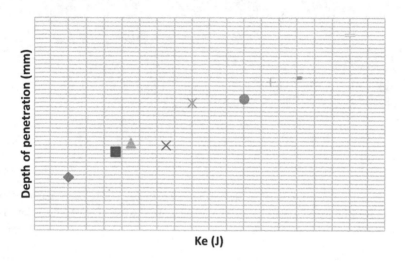

Figure 12. The kinetic energy of the impact against a depth of penetration in the material.

4. Finite element modelling

FEM was carried out by using ANSYS package (ANSYS, 2010). The matrix was identified by 10 nodded tetrahedron continuum elements called SOLID187 and SOLID186 in ANSYS, while for the microspheres ceramic walls SURF154 was used.

In Ansys/ workbench, the material was defined by inputting its mechanical properties that had been found by the static compression test. To define a material in Ansys/workbench, physical properties, linear elastic and plasticity of the material must be defined. The material must have a valid density defined for explicit or implicit simulation. Isotropic elasticity was used to define linear elastic material behaviour by defining Young's modulus and Poisson's ratio. However, the plastic deformation was computed by reference to Von Mises yield criterion. The multi-linear isotropic model was used to define the yield stress (σ_y) as a piece-wise linear function of plastic strain, ε_p. The multi-linear isotropic model has been defined by introducing up to ten of stress-strain pairs.

A load of 100kN was imparted to aluminium matrix syntactic foam that had a dimension of 20 mm x 20mm x20mm. FEA is used to understand the stress distribution around the microspheres ceramics. In addition, the effect of the size of the microspheres and its volume fraction on the composite properties has been studied. Boundary condition was applied to the bottom of the sample that allowed the material to move freely in X and Z directions while Y-direction was fixed. The sample material had been defined as multi-linear isotropic hardening model.

It is found that, the Young's modulus increases with relative wall thickness, which means the particles wall thickness can be used to manage the deformation and fracture behaviour of syntactic foam. In addition, the location of the maximum stress is found on the inner surface of the particles as shown in Figure.13.

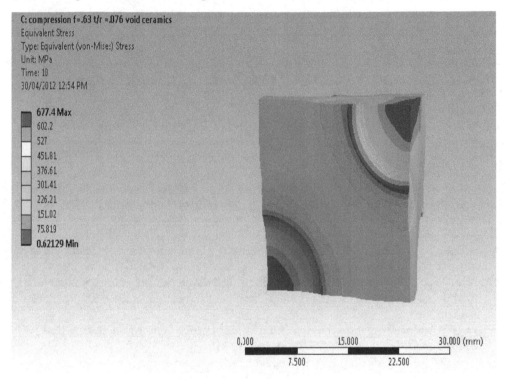

Figure 13. Maximum stress at the internal wall of the particles

5. Summary

- The ability to absorb high energy is the most desirable characteristic of metal matrix synthetic foam materials for use as a protective structure against dynamic loadings.
- The characterisation of the material was done by using quasi-static and dynamic loading. Static compression tests were conducted to find the stress-strain curve. The plateau stress and densification strain are found from those tests which are important properties to determine the energy absorption capability of materials.
- The effect of volume ratio of the metal matrix and the ceramic spheres on the mechanical properties of syntactic foam has, so far, not been thoroughly investigated. Therefore, the different failure modes of metal matrix syntactic foams were observed in previous studies.

- The previous studies suggest that the simplest way to understand the failure mechanism of metal matrix syntactic foam is by comparing the failure modes of closed cell foam, in spite of the difference in the structure of the metal matrix syntactic foam that has ceramics instead of voids.

Author details

M. Altenaiji, G.K. Schleyer and Y.Y. Zhao

University of Liverpool Impact Research Centre, School of Engineering, Liverpool, UK

Acknowledgement

The authors would like to thank United Arab Emirates Government for sponsorship of this project, and also thank Mr Steve Pennington and Mr Rafael Santiago for their support during the experimental work and Dr L.P. Zhang for her assistance in preparing the syntactic foam samples, Dr. Ahmad Sheikh Rafi and Mrs Mariya Krates for her help.

6. References

Arnaud Fazekas, Luc Salvo, Remy Dendievel, Souhail Yossef, Peter Cloetenss, Jean Michel Letang, X-ray tomography applied to the characterization of cellular materials, related finite element modelling problems, Composite science and technology, Vol. 63, 2003, pages 2431-2443.

Balch, D.K. & Dunand, D.C., Load partitioning in aluminium syntactic foams containing ceramic micro-spheres, Acta Materialia, Vol.54, 2006, pp. 1501-1511.

Beguefin, P. and Barbezat, M. 'Caracterisation mecanique des polymeres et composites l ' aide d'une machine d'essais rapide s, Proc 5th Journ~e Nationale DYMAT, Bordeaux, France, 1989

Cantwell W., Morton J. Impact perforation of carbon fibre reinforced plastic, Composites science and technology, Vol.38, 1990, pp. 119-141.

Cantwell W.J. and Morton J., The impact resistance of composite materials –review, Composites, Vol. 22, Number 5, SEPT. 1991, pp.347-362.

Chen W., Lu F., Frew D., Forrestal M. Dynamic Compression Testing of Soft Materials, Journal of Applied Mechanics, Vol.69, 2002, pp. 214-223.

Dorian K. Balch, John G. O'Dwyer, Graham R. Davis, Carl M. Cady, George T. Gray III, David C. Dunand, Plasticity and damage in aluminium syntactic foams deformed under dynamic and quasi-static conditions, Material science & Engineering, Vol.391, 2005, pp. 408-417.

Dorian K. Balch, John G. O'Dwyer, Graham R. Davis, Carl M. Cady, George T. Gray III, David C. Dunand, Plasticity and damage in aluminium syntactic foams deformed under dynamic and quasi-static conditions, Material science & Engineering, Vol.391, 2005, pp. 408-417.

Dou Z.Y., Jiang L.T., Wu G.H., Zhang Q., Xiu Z.Y. and Chen G.Q. High strain rate compression of ceno-sphere-pure aluminium syntactic foams, Scripta Materialia, vol. 57, 2007, pp. 945–948

Dung L., Gupta N., Rohatgi P. The high strain rate compressive response of Mg-Al alloy/ fly ash ceno-sphere composites, JOM, 2011, pp. 48-52.

Fanjing. Y, Geometrical effects in the impact response of composite structures, PhD thesis, University of Liverpool, 2010.

Gibson L.J. Mechanical behaviour of metallic foams, Mater Science, Vol.30, 2000, pp. 191-227.

Goldsmith W, Sackman JL. An experimental study of energy absorption in impact on sandwich plates, International Journal of Impact Engineering, Vol.12, Part(2), 1992, Pp. 241–262.

Gray III, G. T., Classic split-Hopkinson pressure bar testing, in ASM Handbook.Vol. 8: Mechanical Testing and Evaluation, ASM International, Materials Park, Ohio, 2000, pp. 462-476.

Hanssen AG, Enstock L, Langseth M. Close-range blast loading of aluminium panels. Int. J. Impact Eng., Vol. 27, 2002, pp. 593-618.

Harding J., Wood E.O., Campbell J.D., Tensile testing of materials at impact rates of strain, journal of Mechanical Eng. Sci., Vol.2, 1960, pp. 88-96.

Hazell P., Appleby-Thomas G., Herlarr K., Painter J. Inelastic deformation and failure of tungsten carbide under ballistic loading conditions, Materials science and engineering A, Vol.527, 2010, pp.7638-7645.

Hsiao, H. M., Daniel, I. M. and Cordes, R. D., Dynamic compressive behaviour of thick composite materials, Experimental Mechanics, Vol.38- 3, pp. 172-180, 1998.

Hsiao, H. M., Daniel, Strain rate behaviour of composite materials, Composites, part B 29B, 1998, pp 521-533.

I. Irausquín, F. Teixeira-Dias, V. Miranda, J.L. Pérez-Castellanos, Numerical modelling of the dynamic compression of a closed-cell aluminium foam, Universidad Carlos III de Madrid, 28911 Leganés, España.

Kiernan S., Cui L., Gilchrist M., Propagation of a stress wave through a virtual functionally graded foam, International Journal of Non-Linear Mechanics, vol. 44, 2009, pp. 456-468.

Kiser M., He M.Y. and Zok F. W., the mechanical response of ceramic micro-balloon reinforced aluminium matrix composites under compressive loading. Actra Mater, vol. 47, No.9, 1999, pp.2685-2694.

Kohnke P. ANSYS, Inc., 'Theory reference'. ANSYS Academic research-release 13.0, 2010.

Lewis J.L. and Goldsmith W., A biaxial split Hopkinson bar for simultaneous torsion and compression, Rev. Sci. Instrum., Vol.44, 1973, pp.811-813.

Lopatnikov S., Gama B., Haque M.J., Krauthauser C., Gillespie J., Guden M., Hall I., Dynamics of metal foam deformation during Taylor cylinder–Hopkinson bar impact experiment, Composite Structures, vol.61, 2003, pp. 61–71.

Lora j, Gibson and Michael F. Ashby (1997). Cellular solids structure and properties. 2nd edition. Cambridge, UK: Cambridge University press.

M. Altenaiji, G.K. Schleyer, Y.Y. Zhao. Characterisation of aluminium matrix syntactic foams under static and dynamic loading. Applied Mechanics and Materials Vol. 82 (2011) pp 142-147.

N. Gupta, E. Woldesenbet, Micro balloon Wall Thickness Effects on Properties of Syntactic Foams, Journal of Cellular Plastics , Vol. 40, 2004, pp. 461-480.

Novak R.C. and De Crescente M.A., in composite Materials: testing and Design (second conference), American Society for Testing and Materials, Vol. 497, 1972, pp 311-323.

Palmer, R.A., Gao, K., Doan, T.M., Green, L. & Cavallaro, Pressure infiltrated syntactic foams process development and mechanical properties, Material science and Engineering, Vol. 464, no.1-2, 2007, pp.358-366.

Rajendran R., Prem K, Basu S., B. Chandrasekar, A. Gokhale, Preliminary investigation of aluminium foam as an energy absorber for nuclear transportation cask, Materials and design, Vol. 29, 2008, pp. 1732-1739.

Roger L. Ellis (1996). Ballistic impact resistance of graphite epoxy composites with shape memory alloy and extended chain polyethylene spectra™ hybrid component, MSC, Faculty of the Virginia Polytechnic Institute and State, Blacksburg, Virginia.

Rohatgi P.K., Kim J.K. ,Gupta,Simon Alaraj N., Daoud A. Compressive characteristics of A356/fly ash ceno-sphere composites synthesized by pressure infiltration technique, Composites: Part A 37, 2006, pp. 430–437.

S. Barre, T. Chotard, M. L. Benzeggagh, Comparative study of strain rate effects on mechanical properties of glass fibre- reinforced thermoset matrix composites, Composites, Part A - 27A, 1996, pp 1169-1181.

Tao X.F. and Zhao Y.Y. Compressive behaviour of Al matrix syntactic foams toughened with Al particles, Scripta Materialia, vol. 61, 2009, pp. 461–464.

Taylor G.I., The use of flat ended projectiles for determining yield stress, part I: theoretical considerations, proc. R. Soc. (London) A, vol. 194, 1948, pp.289-299.

Vendra, L.J & Rabiei, A study on aluminium-steel composite metal foam processed by casting, Materials Science and Engineering, Vol. 465, no. 1-2, 2007, pp. 59-67.

Wu G.H., Dou Z.Y., Sun D.L., Jiang L.T. , Ding B.S. and He B.F., Compression behaviours of ceno-sphere–pure aluminium syntactic foams, Scripta Materialia, Vol. 56, 2007, pp.221–224.

Xingfu Tao, Fabrication and mechanical properties of metal matrix syntactic foams, Liverpool University, PhD thesis, 2010.

Yiou S., High performance sandwich structures based on novel metal core, PhD thesis, University of Liverpool, 2009

Zhang L.P.and Zhao Y.Y. Mechanical Response of Al Matrix Syntactic Foams Produced by Pressure Infiltration casting, Journal of Composite Materials , Vol. 41, 2007.

Zhao H., El-Nasri I., Abdennadher S., An experimental study on the behaviour under impact loading of metallic cellular materials, International Journal of Mechanical Sciences, vol. 47, 2005, pp. 757–774.

Zhao H., Gary G., Klepaczco J., on the use of a visco-elastic split Hopkinson pressure bar, International Journal of Impact Engineering, vol. 19, 1997, pp. 319-330.

Zhao Y. Y. and Tao X. F., Behaviour of metal matrix under compression, in: Proceedings of Materials Science and Technology 2009, pp. 1785-1794.

Permissions

The contributors of this book come from diverse backgrounds, making this book a truly international effort. This book will bring forth new frontiers with its revolutionizing research information and detailed analysis of the nascent developments around the world.

We would like to thank Ning Hu, Ph.D., for lending his expertise to make the book truly unique. He has played a crucial role in the development of this book. Without his invaluable contribution this book wouldn't have been possible. He has made vital efforts to compile up to date information on the varied aspects of this subject to make this book a valuable addition to the collection of many professionals and students.

This book was conceptualized with the vision of imparting up-to-date information and advanced data in this field. To ensure the same, a matchless editorial board was set up. Every individual on the board went through rigorous rounds of assessment to prove their worth. After which they invested a large part of their time researching and compiling the most relevant data for our readers. Conferences and sessions were held from time to time between the editorial board and the contributing authors to present the data in the most comprehensible form. The editorial team has worked tirelessly to provide valuable and valid information to help people across the globe.

Every chapter published in this book has been scrutinized by our experts. Their significance has been extensively debated. The topics covered herein carry significant findings which will fuel the growth of the discipline. They may even be implemented as practical applications or may be referred to as a beginning point for another development. Chapters in this book were first published by InTech; hereby published with permission under the Creative Commons Attribution License or equivalent.

The editorial board has been involved in producing this book since its inception. They have spent rigorous hours researching and exploring the diverse topics which have resulted in the successful publishing of this book. They have passed on their knowledge of decades through this book. To expedite this challenging task, the publisher supported the team at every step. A small team of assistant editors was also appointed to further simplify the editing procedure and attain best results for the readers.

Our editorial team has been hand-picked from every corner of the world. Their multi-ethnicity adds dynamic inputs to the discussions which result in innovative outcomes. These outcomes are then further discussed with the researchers and contributors who give their valuable feedback and opinion regarding the same. The feedback is then collaborated with the researches and they are edited in a comprehensive manner to aid the understanding of the subject.

Apart from the editorial board, the designing team has also invested a significant amount of their time in understanding the subject and creating the most relevant covers. They scrutinized every image to scout for the most suitable representation of the subject and create an appropriate cover for the book.

The publishing team has been involved in this book since its early stages. They were actively engaged in every process, be it collecting the data, connecting with the contributors or procuring relevant information. The team has been an ardent support to the editorial, designing and production team. Their endless efforts to recruit the best for this project, has resulted in the accomplishment of this book. They are a veteran in the field of academics and their pool of knowledge is as vast as their experience in printing. Their expertise and guidance has proved useful at every step. Their uncompromising quality standards have made this book an exceptional effort. Their encouragement from time to time has been an inspiration for everyone.

The publisher and the editorial board hope that this book will prove to be a valuable piece of knowledge for researchers, students, practitioners and scholars across the globe.

List of Contributors

Ali Hammood and Zainab Radeef
Department of Materials Engineering-University of Kufa, Iraq

Dewan Muhammad Nuruzzaman and Mohammad Asaduzzaman Chowdhury
Department of Mechanical Engineering, Dhaka University of Engineering and Technology (DUET), Gazipur, Bangladesh

Konstantin N. Rozanov
Institute for Theoretical and Applied Electromagnetics, Russian Academy of Sci., Moscow, Russia

Marina Y. Koledintseva
Missouri University of Science and Technology, Rolla, MO, USA

Eugene P. Yelsukov
Physical and Technical Institute, Ural Branch of Russian Academy of Sci., Izhevsk, Russia

Pavel Koštial and Ivan Ružiak
VŠB – Technical University of Ostrava, Faculty of Metallurgy and Material Engineering, Department of Material Engineering, Ostrava, Czech Republic

Jan Krmela
University of Pardubice, Jan Perner Transport Faculty, Department of Transport Means and Diagnostics, Pardubice, Czech Republic

Karel Frydrýšek
ING-PAED IGIP, VŠB – Technical University of Ostrava, Faculty of Mechanical Engineering, Department of Mechanics of Materials, Ostrava, Czech Republic

Jinxiang Chen and Juan Xie
International Institute for Urban Systems Engineering & School of Civil Engineering, Southeast University, China

Qing-Qing Ni
Dept. of Functional Machinery & Mechanics, Shinshu University, Japan

Rafic Younes
LISV, University of Versailles Saint-Quentin, Versailles, France
Faculty of Engineering, Lebanese University, Rafic Hariri campus, Beirut, Lebanon

Ali Hallal
LISV, University of Versailles Saint-Quentin, Versailles, France
L3M2S, Lebanese University, Rafic Hariri campus, Beirut, Lebanon

Farouk Fardoun
L3M2S, Lebanese University, Rafic Hariri campus, Beirut, Lebanon

Fadi Hajj Chehade
L3M2S, Lebanese University, Rafic Hariri campus, Beirut, Lebanon

M. Sayuti
Department of Mechanical and Manufacturing Engineering, Faculty of Engineering, Universiti Putra Malaysia, Serdang, Selangor, Malaysia
Department of Industrial Engineering, Faculty of Engineering, Malikussaleh University, Lhokseumawe, Aceh - Indonesia

S. Sulaiman, B.T.H.T. Baharudin and M.K.A. Arifin
Department of Mechanical and Manufacturing Engineering, Faculty of Engineering, Universiti Putra Malaysia, Serdang, Selangor, Malaysia

T.R. Vijayaram
Faculty of Engineering and Technology (FET) Multimedia University, Jalan Ayer Keroh Lama, Bukit Beruang, Melaka, Malaysia

Nor Bahiyah Baba
TATI University College (TATIUC), Terengganu, Malaysia

Go Yamamoto and Toshiyuki Hashida
Fracture and Reliability Research Institute (FRRI), Tohoku University, Japan

M. Altenaiji, G.K. Schleyer and Y.Y. Zhao
University of Liverpool Impact Research Centre, School of Engineering, Liverpool, UK